CONTROLS AND ASSURANCE IN THE CLOUD:

Using COBIT® 5

ISACA®

With more than 110,000 constituents in 180 countries, ISACA (*www.isaca.org*) helps business and IT leaders maximize value and manage risk related to information and technology. Founded in 1969, the nonprofit, independent ISACA is an advocate for professionals involved in information security, assurance, risk management and governance. These professionals rely on ISACA as the trusted source for information and technology knowledge, community, standards and certification. The association, which has 200 chapters worldwide, advances and validates business-critical skills and knowledge through the globally respected Certified Information Systems Auditor® (CISA®), Certified Information Security Manager® (CISM®), Certified in the Governance of Enterprise IT® (CGEIT®) and Certified in Risk and Information Systems Control™ (CRISC™) credentials. ISACA also developed and continually updates COBIT®, a business framework that helps enterprises in all industries and geographies govern and manage their information and technology.

Disclaimer

ISACA has designed and created *Controls and Assurance in the Cloud: Using COBT® 5* (the "Work') primarily as an educational resource for assurance professionals. ISACA makes no claim that use of any of the Work will assure a successful outcome. The Work should not be considered inclusive of all proper information, procedures and tests or exclusive of other information, procedures and tests that are reasonably directed to obtaining the same results. In determining the propriety of any specific information, procedure or test, assurance professionals should apply their own professional judgment to the specific circumstances presented by the particular systems or information technology environment.

ISACA
3701 Algonquin Road, Suite 1010
Rolling Meadows, IL 60008 USA
Phone: +1.847.253.1545
Fax: +1.847.253.1443
Email: *info@isaca.org*
Web site: *www.isaca.org*

Provide feedback: *www.isaca.org/controls-and-assurance-in-the-cloud*
Participate in the ISACA Knowledge Center: *www.isaca.org/knowledge-center*
Follow ISACA on Twitter: *https://twitter.com/ISACANews*
Join ISACA on LinkedIn: ISACA (Official), *http://linkd.in/ISACAOfficial*
Like ISACA on Facebook: *www.facebook.com/ISACAHQ*

ISBN 978-1-60420-464-3
Controls and Assurance in the Cloud: Using COBT® 5

Acknowledgments

ISACA wishes to recognize:

Development Team
Chris Kappler, CISSP, CSSK, PwC Belgium, Belgium
Bart Peeters, CISA, PwC Belgium, Belgium
Dirk Van Droogenbroeck, CISA, ISO 27001 Lead Auditor, ITIL V3, TOGAF, PwC Belgium, Belgium
Sven Van Hoorebeeck, CISA, ITIL V3, TOGAF, PwC Belgium, Belgium

Work Group
Phil J. Lageschulte, CGEIT, CPA, KPMG LLP, USA
Nnamdi Nwosu, CISA, CSTE, CSQA, ITIL, PMP, Moleworth Consulting, Nigeria
Antonio Ramos Garcia, CISA, CISM, CRISC, Leet Security and Nplus 1 Intelligence & Research, Spain
Anne Maria Yrjana, CISA, CRISC, Tieto Finland, Finland
Nikolaos Zacharopoulos, CISA, CISSP, DeutschePost-DHL, Germany

Expert Reviewers
Jeimy J. Cano, Ph.D., CFE, CMAS, Ecopetrol, Colombia
P. W. Carey, CISA, CISSP, Compliance Partners, LLC, Barrington, IL, USA
Abhik Chaudhuri, PMP, India
Epsilon Ip, CISA, CISM, CRISC, CISSP, ISSMP, ISSAP, Cathay Pacific, Hong Kong
John Jasinski, CISA, CGEIT, SSBB, ITIL Expert, ISO20K, ITSMBP, USA
Ramaswami Karunanithi, CISA, CGEIT, CRISC, CIA, CPA, NSW Govt. Dept of Ageing, Disability & Home Care, Australia
Lily M. Shue, CISA, CISM, CGEIT, CRISC, LMS Associates, LLP, USA
Theodoros Stergiou, Ph.D., CPMM, CCDA, CSSDS, Intracom Telecom, Greece
Daniel Zimerman, CISA, CRISC, CISSP, IQ Solutions, Inc., USA

ISACA Board of Directors
Tony Hayes, CGEIT, AFCHSE, CHE, FACS, FCPA, FIIA, Queensland Government, Australia, International President
Allan Boardman, CISA, CISM, CGEIT, CRISC, ACA, CA (SA), CISSP, Morgan Stanley, UK, Vice President
Juan Luis Carselle, CISA, CGEIT, CRISC, RadioShack Mexico, Mexico, Vice President
Ramses Gallego, CISM, CGEIT, CCSK, CISSP, SCPM, Six Sigma Black Belt, Dell, Spain, Vice President
Theresa Grafenstine, CISA, CGEIT, CRISC, CGAP, CGMA, CIA, CPA, US House of Representatives, USA, Vice President
Vittal Raj, CISA, CISM, CGEIT, CFE. CIA, CISSP, FCA, Kumar & Raj, India, Vice President
Jeff Spivey, CRISC, CPP, PSP, Security Risk Management Inc., USA, Vice President
Marc Vael, Ph.D., CISA, CISM, CGEIT, CRISC, CISSP, Valuendo, Belgium, Vice President
Gregory T. Grocholski, CISA, The Dow Chemical Co., USA, Past International President
Kenneth L. Vander Wal, CISA, CPA, Ernst & Young LLP (retired), USA, Past International President
Christos K. Dimitriadis, Ph.D., CISA, CISM, CRISC, INTRALOT S.A., Greece, Director
Krysten McCabe, CISA, The Home Depot, USA, Director
Jo Stewart-Rattray, CISA, CISM, CGEIT, CRISC, CSEPS, BRM Holdich, Australia, Director

Knowledge Board
Christos K. Dimitriadis, Ph.D., CISA, CISM, CRISC, INTRALOT S.A., Greece, Chairman
Rosemary M. Amato, CISA, CMA, CPA, Deloitte Touche Tohmatsu Ltd., The Netherlands
Steven A. Babb, CGEIT, CRISC, Vodafone, UK
Thomas E. Borton, CISA, CISM, CRISC, CISSP, Cost Plus, USA
Phil J. Lageschulte, CGEIT, CPA, KPMG LLP, USA
Anthony P. Noble, CISA, Viacom, USA
Jamie Pasfield, CGEIT, ITIL V3, MSP, PRINCE2, Pfizer, UK

Acknowledgments *(cont.)*

Guidance and Practices Committee
Phil J. Lageschulte, CGEIT, CPA, KPMG LLP, USA, Chairman
John Jasinski, CISA, CGEIT, ISO20K, ITIL Expert, SSBB, ITSMBP, USA
Yves Marcel Le Roux, CISM, CISSP, CA Technologies, France
Aureo Monteiro Tavares Da Silva, CISM, CGEIT, Brazil
Jotham Nyamari, CISA, Deloitte, USA
James Seaman, CISM, CRISC, A.Inst.IISP, CCP, QSA, RandomStorm, UK
Gurvinder Singh, CISA, CISM, CRISC, Australia
Siang Jun Julia Yeo, CISA, CRISC, CPA (Australia), MasterCard Asia/Pacific Pte. Ltd., Singapore
Nikolaos Zacharopoulos, CISA, CISSP, DeutschePost–DHL, Germany

Table of Contents

Page intentionally left blank

List of Figures

1. Introduction

Cloud computing has evolved into an ever-growing presence in the IT industry and far beyond. Within five years, cloud adoption changed from an idea that often met much resistance to a solution that is growing exponentially.

In many countries, e.g., Australia, government agencies have an explicit obligation to consider cloud services when procuring new information and communication technology (ICT) requirements for their test and development needs, and to migrate public facing web sites to public cloud services. The agencies must choose cloud services when they represent the best value and adequate risk management compared to other available options.

Using cloud services brings multiple benefits to cloud users, but it also raises many concerns, which, if not handled well, can quickly turn the cloud experience into an information-security management nightmare derived from the loss of controls over physical and logical assets.

Using This Publication

The purpose of this publication is to:
- Provide readers with an understanding of cloud computing, its enablers and the business drivers behind this new way to deliver IT services.
- Identify and create awareness about the real-world cloud-related business challenges, risk and possible mitigating actions.
- Provide process practices and frameworks that can be used to address the challenges and maximize value in the cloud.

Readers will not only learn how to understand the cloud computing landscape, but also how to build the relevant controls and governance mechanisms around it.

Cloud computing can produce significant business opportunities; however, the associated information risk should not be underestimated and should be addressed. This publication provides insight into how the COBIT 5 (the framework), *COBIT 5 for Risk*, *COBIT 5 for Assurance* and *COBIT 5 for Information Security* professional guides can assist enterprises in assessing the cloud's business value vs. its business risk. Additionally, this book provides guidance on how to determine whether the risk aligns with the established levels of risk within the enterprise and whether the rewards and benefits are worth the cost and effort to mitigate that risk.

Management should consider the monitoring mechanisms that are appropriate and necessary for the enterprise's specific circumstances. Management may choose not to include all of the activities and approaches discussed in this publication and, similarly, may choose to include activities not mentioned in this publication. In either case, customization of the approaches described in this publication will undoubtedly be necessary to reflect the specific circumstances of each enterprise.

Page intentionally left blank

2. Cloud Computing Fundamentals

Cloud Computing Defined

Cloud computing is defined by the US National Institute of Standards and Technology (NIST) as a:[1]

> *Model for enabling convenient, on-demand network access to a shared pool of configurable computing resources (e.g., networks, servers, storage, applications, and services) that can be rapidly provisioned and released with minimal management effort or service provider interaction.*

Following are definitions for the **five essential characteristics** of the cloud:

- **On-demand self-service**—Computing capabilities can be provisioned without human interaction from the service provider.
- **Broad network access**—Computing capabilities are available over the network and can be accessed by diverse client platforms.
- **Resource pooling**—Computer resources are pooled to support a multitenant model.
- **Rapid elasticity**—Resources can scale up or down rapidly and, in some cases, automatically, in response to business demands.
- **Measured service**—Resource utilization can be optimized by leveraging charge-per-use capabilities.

Adding computer virtualization to the NIST definition (on-demand computer resources requiring minimal management effort) creates a cloud computing model that offers enterprises virtual processing power in a variety of possible implementations (**figure 1**).[2]

Figure 1—Cloud Computing Service Delivery and Deployment Model

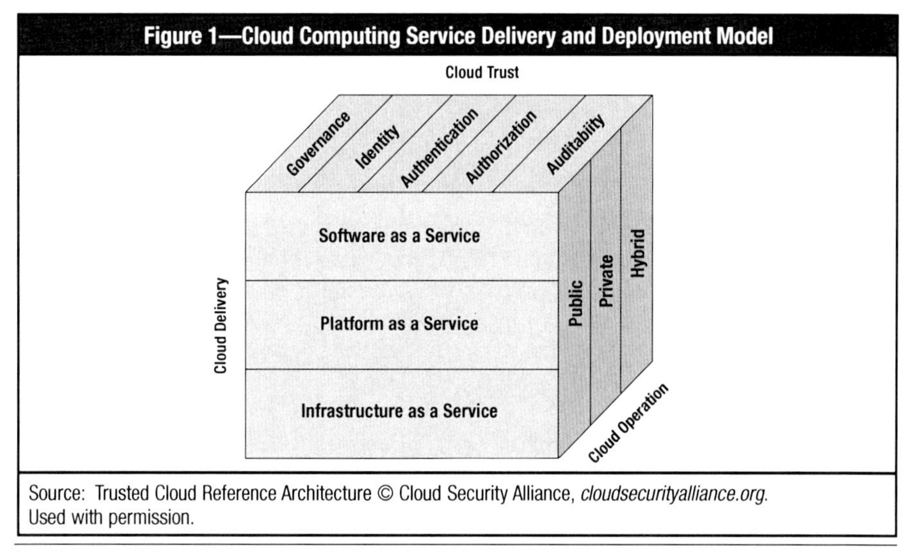

Source: Trusted Cloud Reference Architecture © Cloud Security Alliance, *cloudsecurityalliance.org*. Used with permission.

[1] Mell, Peter; Timothy Grance; US National Institute of Standards and Technology (NIST) Special Publication (SP) 800-145 (Draft), "The NIST Definition of Cloud Computing," NIST, USA, 2011

[2] Cloud Security Alliance, "Trusted Cloud Reference Architecture," 25 February 2013, *https://downloads.cloudsecurityalliance.org/initiatives/tci/TCI_Reference_Architecture_v2.0.pdf*

Cloud Drivers

Cloud computing is viewed as a significant change to the platform in which business services are translated, used and managed. Many consider it to be as large a shift in IT as was the advent of the personal computer (PC) or of Internet access. However, a major difference between the cloud and those technologies is that the introductions of those earlier technologies encompassed a slower development phase. With the cloud, the required pieces have come together more rapidly for implementation. Some of the drivers bringing the cloud to the attention of enterprise decision makers are:

- **Optimized resource utilization**—Enterprises typically use just 15 to 20 percent of server computing resources.[3] This means that they have five times the computing capacity than is typically used. By using a pay-as-you-go cloud solution, resources become available when needed and are liberated when no longer needed; there is near to perfect alignment with actual demand.
- **Cost savings**—Increased server utilization plus the transition of computational capability from acquired and maintained computers to rented cloud services change the computing cost paradigm from a capital expenditure (CAPEX) to an operational expenditure (OPEX), with potentially significant up-front and total cost savings. Indeed, flexible, on-demand services enable solution testing without significant capital investments and provide transparency of usage charges to drive behavioral change within organizations.
- **Better responsiveness**—On-demand, agile, scalable and flexible services that can be implemented quickly provide organizations with the ability to respond to changing requirements and peak periods.
- **Faster cycle of innovation**—By using cloud, innovation is handled a lot faster than when handled within the enterprise. Patch management and upgrades to new versions become more flexible. For the cloud user, upgrading to a new software version is often nothing more than typing a different URL into the web browser.
- **Reduced time for implementation**—Cloud computing provides processing power and data storage as needed and at the capacity needed, in near-real time, not requiring the weeks or months (or CAPEX) that accrue when a new business initiative is brought online in a traditional IT enterprise.
- **Resilience**–A large, highly resilient environment reduces the potential for system failure. The failure of one component of a cloud-based system has less impact on overall service availability and reduces the risk of downtime.

Depending on business needs, any or all of these benefits could be a sufficient reason to consider a cloud computing solution. The recent world economy has pushed many enterprises to be more fiscally conservative. In the IT space, cloud computing presents a potentially significant savings by enabling enterprises to maximize dynamic computing on a pay-per-use basis. By using the governance processes described in chapter 2, this advantage can be leveraged across entire enterprises.

Each cloud opportunity or program has many variables, benefits and risk that affect the decision whether a cloud application should be adopted from a risk/business value standpoint. The enterprise must weigh those variables to decide whether the cloud is an appropriate solution.

---

[3] Kanellos, Michael; "Is Cyber Monday Really Energy Efficient?," Greentech Enterprise, 24 November 2010, *www.greentechmedia.com/articles/read/is-cyber-monday-really-energy-efficient*

The benefits and risk associated with the cloud vary based on the following factors:

- **Type of cloud service model**—Software as a Service (SaaS), Platform as a Service (PaaS) or Infrastructure as a Service (IaaS), and associated service models, such as Security as a Service (SecaaS), Storage as a Service (StaaS), etc. Each cloud service model has varied business purposes and levels of business risk.
- **Robustness of the enterprise's existing IT operations**—Enterprises need to ensure that their current governance, risk management and information security enablers are well defined and managed within the existing IT operations. New threats and vulnerabilities may be identified in the cloud, but if the enterprise is prepared and its current IT operations are able to handle these issues, the overall risk to the enterprise may be lower.
- **Current level of business risk acceptance**—The level of risk an enterprise is willing to accept varies among industries and among enterprises within the same industry.
- **Aggregated "street value" of the data to be promoted to the cloud**—Enterprises need to assess the value of the data promoted to the cloud in terms of the potential value of that data to people with malicious intent.
- **Internal security classification of data being promoted to the cloud**—Additional to the "street value," data have internal value to the enterprise, which provides the enterprise with a vested interest in keeping them proprietary and not releasing data publicly.
- **Identified compliance obligations of the data shared within the cloud**—Personally identifiable information (PII) security controls and financial reporting compliance are two prime examples of compliance obligations that need to be managed in the cloud.
- **Cloud service provider (CSP) risk**—Enterprises must exercise due diligence when considering moving services to the cloud. Because no consistent cloud security standards have been commonly accepted, CSPs may have different approaches to cloud security. CSPs should be following best practices and making use of internationally accepted standards. It is very important for enterprises to have their own well-defined requirements to be able to reap the maximum benefit from the due diligence phase.

The COBIT family of publications provides a reference model, assessment methodology and improvement tools to support management in governing and managing intended cloud benefits and cloud risk, by providing structure, context and common vocabulary.

Cloud Service Models

Cloud computing is implemented in three delivery models: SaaS, PaaS and IaaS (SPI) (see **figure 2**). Each delivery model provides a distinct computing service to the enterprises that use them:

- **IaaS**—Provides online processing or data storage capacity. This cloud service is ideal for enterprises considering very large, one-time processing projects or infrequent, extremely large data storage requirements i.e., test environments. IaaS offers the capability to provision processing, storage, networks and other fundamental computing resources, enabling the customer to deploy and run arbitrary software, which can include operating systems (OSs) and applications.

- **PaaS**—Provides the application development sandbox in the cloud. PaaS provides the capability to deploy customer-created, or –acquired, applications that are developed using programming languages and tools that are offered by the provider. The CSP offers enterprise developers elemental service-oriented architecture (SOA) application building blocks to configure a new business application. In-house development requires development, testing and user acceptance platforms, all separate from the production environment. Through PaaS, enterprise developers can rent their development environment, complete with an SOA tool kit, and the enterprise is charged only for the time the tools and environment are in use.
- **SaaS**—Provides a business application that is used by many individuals or enterprises concurrently. SaaS provides the most used cloud applications to nearly everyone online. G-mail™, Yahoo® user applications, Google Docs and Microsoft® Online Services are all popular consumer-directed SaaS applications; SalesForce.com®, ServiceNow® and WorkDay® are popular business-directed SaaS applications. SaaS allows customers to use the provider's applications, which are running on a cloud infrastructure. The applications are accessible from various client devices through a thin client interface, such as a web browser.

Figure 2—Cloud Delivery Models

Service Model[4]	Description	Considerations
IaaS	Capability to provision processing, storage, networks and other fundamental computing resources, offering the customer the ability to deploy and run arbitrary software, which can include OSs and applications. IaaS puts these IT operations into the hands of a third party.	IaaS can provide infrastructure services such as servers, disk space, network devices and memory and is designed for users wanting complete freedom with regard to the OS and applications they use.
PaaS	Capability to deploy onto the cloud infrastructure customer-created or customer-acquired applications developed using programming languages and tools supported by the provider.	PaaS provides an application development sandbox and is specifically designed for developers.
SaaS	Capability to use the provider's applications running on cloud infrastructure. The applications are accessible from various client devices through a thin client interface such as a web browser (e.g., web-based email).	SaaS provides applications that are complete and available on demand to the end customer. Traditional licensing and asset management are changed.

Cloud Deployment Models

The three cloud service delivery models are offered to cloud customers in four cloud deployment models: private, public, community and hybrid (see **figure 3**).

- **Private cloud**—One client (one individual or one enterprise). Several different departments or divisions may be represented, but all exist within the same enterprise. Private clouds often employ virtualization within an enterprise's existing computer

[4] Pijanowski, Keith; "Understanding Public Clouds: IaaS, PaaS and SaaS," Keith Pijanowski's Blog, 31 May 2009, *www.keithpij.com/Home/tabid/36/EntryID/27/Default.aspx*

servers to improve computer utilization. A private cloud also typically involves provisioning and metering components, enabling rapid deployment and chargeback where appropriate. This model is most closely related to traditional IT outsourcing models in the marketplace, but can also be an enterprise's internal delivery model.

- **Public cloud**—An offering from one CSP to many clients who share the cloud processing power and storage capacity concurrently. Public cloud clients share applications, processing power and data storage space communally. Client data are commingled, but segregated, for example, through the use of metatags.
- **Community cloud**—A private-public cloud with clients who have a common connection or affiliation, such as a trade association, industry or locality. The community cloud business model allows a CSP to provide cloud tools and applications that are specific to the needs of the community. When the community uses a PaaS cloud, the SOA applets can be specific to communal requirements, e.g., business-process-specific, industry-specific.
- **Hybrid cloud**—A combination of two or more of the previously defined deployment models. Each of the three cloud deployment models has specific advantages and disadvantages that are relative to the other deployment models. A hybrid cloud leverages the advantage of the other cloud models, providing a more optimal user experience.

Within the matrix of cloud delivery/deployment variants, a private cloud deployment of any delivery model is the most similar to traditional IT enterprises and, thus, offers the least amount of new risk and security challenges. Any public cloud deployment, especially SaaS with a significant number of concurrent users, presents the greatest assurance challenges for security and risk managers. **Figure 3** shows the relationship between the service and deployment models and their cumulative risk.

Figure 3—Cloud Computing Risk Map

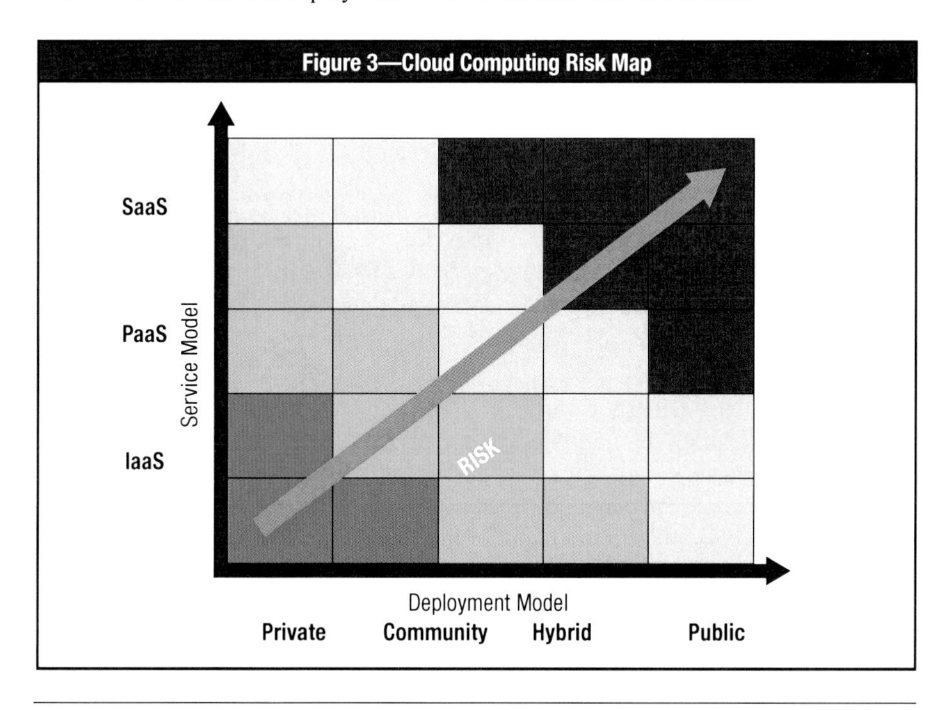

Figure 4 summarizes the available cloud deployment models.

Figure 4—Cloud Deployment Models	
Deployment Model	**Description**
Private cloud	• Operated solely for one enterprise • May be managed by the enterprise or a third party • May exist on- or off-premise
Public cloud	• Made available to the general public or a large industry group • Owned by an organization selling cloud services
Community cloud	• Shared by several enterprises • Supports a specific community that has a shared mission or interest • May be managed by the enterprises or a third party • May reside on- or off-premise
Hybrid cloud	A combination of two or more cloud deployment models (private, community or public) that remain unique entities, but are bound together by standardized or proprietary technology that enables data and application portability, e.g., cloud bursting for load balancing between clouds

Levels of information security vary among the private, public and community deployment models, with private clouds having the most restricted user access and, most likely, the least exposure to threats.

Likewise, the costs of services vary across the different deployment models. Private cloud services are currently the most costly option; public clouds are the least costly. For users who want to save on expenses, the hybrid cloud offers a combination of two or more deployment models with varying levels of security, as needed. Users can choose to leverage private or community clouds for their most business-critical data while choosing to utilize the public cloud for data that are already publicly available or for other nonclassified data or applications. Some enterprises may just accept the risk and choose to use the public cloud regardless of data classification. The decision on how to leverage the cloud is unique for each enterprise.

Because of the dynamic and evolving nature of this industry and the currently limited acceptance of standards or security certifications, offerings of CSPs are not standardized. It is the responsibility of prospective cloud clients to determine the amount of security provisioning that they will require depending on the type of application and the security classifications of the data they plan to promote to the cloud.

Evolution of the Cloud

New Service Offerings

Cloud service offerings are constantly undergoing major changes and evolving into new and better service offerings. Within the three "traditional" service deliveries (IaaS, PaaS and SaaS), a variety of new services are emerging that are closely related. Most of the new services focus on replacing the traditional in-house IT services with cloud variants and on the activities with which internal IT departments frequently struggle.

These services benefit an enterprise in multiple ways:

- Easier access to new technologies. Smaller enterprises gain access to a wide range of solutions that were previously beyond their financial reach.
- Technical knowledge of CSP and access to newer technology that can provide competitive edge through the services delivered
- Correct and structured implementation of the targeted services

The adoption of cloud technologies should always be subject to careful study and should be aligned and integrated with the internal processes and procedures of an enterprise.

The following services are rising in importance:

- **Security as a Service (SecaaS)**—SecaaS comes in two major forms:
  - The CSP provides standalone managed security services ranging from antivirus scanning and mail security to full deployments of end-point security.
  - The CSP offloads appliance utilization for the client, and CPU- and memory-intensive activities are moved to cloud services. For example, antivirus activities on unified threat management (UTM) devices are often offloaded to a SecaaS provider to reduce the number of chassis at the client site. The advantage to the client is minimized risk when applying patches or updates, because they are no longer directly linked to the device.

  Note: The Cloud Security Alliance (CSA) is a nonprofit organization with a mission to promote best practices for providing security assurance with cloud computing and can be considered a main source of information on this topic.
- **Disaster Recovery as a Service (DRaaS)**—The CSP offers its cloud infrastructure to provide an enterprise with a disaster recovery (DR) solution. In most cases, the CSP not only provides back-up equipment and storage, but also provides services for a business continuity plan (BCP), if it is not yet available. The benefits for DRaaS include:
  - The cost for an in-house DR infrastructure is reduced significantly. Because DR is often considered to be a necessity rather than core business, the return on investment (ROI) in DR services can be significant.
  - Offsite storage means that the DR environment is less likely to fail in the case of a major disaster.
- **Identity as a Service (IDaaS)**—IDaaS is a relatively new cloud service and currently has two interpretations:
  - The management of identities in the cloud that is separated from the users and applications that use the identities. This can be either managed identity services, including provisioning, or management for both onsite and offsite services. Delivering a single sign-on (SSO) solution can also be part of the cloud service offering.
  - The delivery of an identity and access management (IAM) solution. IDaaS is often a hybrid solution where access and roles are configured by the CSP and users are authorized by enterprise internal solutions. This is known as a federated model.

- **Data Storage and Data Analytics as a Service (Big Data)**—Big data is the next step in data analysis that makes it possible to analyze all types of data by taking away the constraints on volume, variety, velocity and veracity. These constraints are not taken away by new big data technologies; they are, rather, removed through a synergy between new technologies and the extended capabilities provided by cloud computing. Limitless volume availability and variety allows enterprises to re-use their "old" data for new purposes. Furthermore, big data technology facilitates the ability of enterprises to find patterns in their current data, which influences their way of doing business. In addition to the advice of experienced people, enterprises receive decision-making support from the information that results from big data analysis, such as real-time reporting and predictive analysis.
- **Information as a Service (InfoaaS)**—This service builds on the big data concept—rather than providing the raw data or the algorithms that are used for trending, InfoaaS provides the required information. With this new service, the result of a query is more important than the query itself.
- **Integration Platform as a Service (IPaaS)**—IPaaS, which is also called "cloud integrator", is defined by Gartner as "a suite of cloud services enabling development, execution and governance of integration flows connecting any combination of on premises and cloud-based processes, services, applications and data within individual or across multiple organizations."[5] Many enterprises are implementing a hybrid model in which some of their data, applications, services and infrastructure are maintained locally on site, while others are provisioned by a cloud provider. Integrating all these business resources can be a complex mission. Cloud integrators can help a business cope with this complexity and facilitate the integration without the need to constantly modify and maintain diverse and often incompatible applications. The main advantage is that IPaaS enables efficient and cost-saving methods to ensure IT integration throughout the enterprise. Furthermore, IPaaS provides a more robust solution in the areas of data confidentiality, integrity and availability and data governance, risk and compliance.
- **Forensics as a Service (FRaaS)**—A relatively newer service that "establishes a cloud forensic investigative process, which can be implemented within a cloud ecosystem, integrated with tolls that should endure relevant information gathered, verified and stored in a manner that is forensically sound and legally defensible."[6]

Cloud Service Broker

As the adoption of cloud computing gained importance, a new kind of service provider emerged: the cloud service broker (CSB). A CSB is a third-party provider with access to multiple data centers and cloud service offerings. A CSB integrates and tailors those various services into one service that can be intertwined with enterprise in-house applications and systems.

Advantages of a CSB:

- A CSB helps an enterprise to determine the best possible framework for integration with cloud services.

---

[5] *www.gartner.com/it-glossary/information-platform-as-a-service-ipaas*

[6] Rav Gagan Shende, Jon; "Forensics as a Service," EBSL Technologies International, USA, 2013, *www.igi-global.com/chapter/forensics-service/73966*

- Because of the integration with various cloud servers, a CSB can guarantee interoperability between the various cloud services needed. Additional support for in-house applications ensures that the cloud adventure integrates seamlessly within the enterprise environment.
- Because of the integration with various CSPs, going to the cloud becomes cost-effective.

However, for CSBs to be successful, an important prerequisite must be fulfilled: cloud computing standards to increase benefits and reduce risk. Also, cloud standardization will help CSPs and users build more robust and portable solutions.

Cloud Standardization

Moving to the cloud still provides many challenges for enterprises to overcome. Portability, interoperability, certifications, correct service level agreements (SLAs) and, in general, knowing what to expect are only a few of the challenges. Hence, governments are trying to encourage cloud standardization and are working on developing legislation concerning cloud computing. The European Community, for example, is working on various cloud initiatives by using working groups, named Cloud Select Industry Groups (C-SIGs). The most important initiatives that the C-SIGs are currently working on are a cloud code of conduct, a voluntary datacenter certification, standardized SLAs and data protection.

Standardization is needed when writing programming code and developing applications. Achieving a high degree of standardization in these tasks will allow or increase portability opportunities. Portability is important in cloud computing, because the goal is to use the same applications in different environments.

Certifications in cloud operations, security or code development are not only important for enterprises, but also for individuals. Cloud-member organizations are starting initiatives to provide their members with the opportunity for certification in cloud computing. For example, the Cloud Security Alliance launched the Certificate of Cloud Security Knowledge (CCSK), allowing its members to demonstrate and certify their knowledge of the security aspects involved in cloud computing.

G-Cloud

Governments are becoming more involved and prone to use the cloud. Public-sector IT investments are increasingly influenced by financial constraints, rapidly aging technology and a higher standard of service delivery that is demanded by the community. Cloud services have the potential to address these challenges by improving the agility, scalability and reliability of IT services and providing the agility to respond to changing business needs.

Government cloud (G-Cloud) is government's answer to this new approach of IT sourcing and management and is considered critical to achieve value, decrease the cost of its IT infrastructure, drive innovation and support sustainable investments, while increasing the flexibility and operability between agencies. Government's shift to a service orientation is taking advantage of the increasing commoditization of IT and the rapidly developing cloud-computing industry. Furthermore, G-Cloud helps a government to become more environmentally friendly, because a joint cloud infrastructure significantly decreases power requirements and carbon exhaust.

One of the most developed G-Clouds is in the United Kingdom. Other countries, such as Canada, are starting to develop their own government clouds. The Australian Federal government and the various state governments are aggressively pursuing the cloud option, because they believe that they can "achieve greater efficiency, generate greater value from ICT investment, deliver better services and support a more flexible workforce."

A second evolution in the G-Cloud is its deployment to citizens. In this case, the G-Cloud is being used as a cloud broker service to supply various cloud solutions to citizens.

Data Protection

One of the most difficult challenges for both a cloud user and CSP is how to protect data. Especially in regard to privacy, many hurdles need to be overcome. The necessity is shown by the emergence of the G-Cloud and many initiatives on a country or regional level. National laws increase the difficulty; for example, France and Luxembourg have incorporated in their laws the stipulation that data can never leave the country. This challenge is also valid for banks and national security information.

Often, the data protection issues that pertain to larger, well-established CSPs do not pertain to smaller CSPs with only local and in-country data sites and provide new opportunities for the smaller CSPs. For increased efficiency, larger CSPs frequently have centralized data centers in another part of the world.

Chapter 5, Assurance in Cloud Computing, discusses how frameworks such as COBIT 5 and the Cloud Control Matrix and ISO 27001 can be used to create a standard way to assess cloud data security.

Cloud Computing Challenges

The benefits of cloud computing also incorporate unique and notable technical and business challenges. Following are the most relevant cloud computing business challenges in relation to the seven COBIT 5 enablers:

1. Principles, Policies and Frameworks:

- *Cloud security policy/procedure transparency*—Some CSPs may have less transparency than others when it comes to their current information security policies. The rationalization for this is that the policies may be proprietary. This practice may cause conflict with clients' information compliance requirements. Clients must make sure that they understand detailed contracts with SLAs and that those contracts provide the desired level of security to ensure that CSPs are applying appropriate controls.

- *Compliance requirements*—For the many compliance requirements—including privacy and PII laws, Payment Card Industry (PCI) requirements and various financial reporting laws—today's cloud computing services can challenge various compliance audit requirements currently in place. Data location, cloud computing security policy transparency and IAM are all challenging issues in compliance audit efforts.

2. **Processes:**
   - *Adequate security controls*—Consideration should be given to whether the CSP has adequate security controls/detection systems in use, e.g., penetration detection. If such a system is in use, it is important to ensure that it has the required sophistication to monitor all cloud computing activities adequately. It is also important to consider whether a real-time digital dashboard is provided to user managers, along with audit logs and records of security incidents.
3. **Organisational Structures:**
   - *Public cloud server owners' due diligence*—Trust is a major component in the cloud computing business model. When contemplating transferring critical organizational data to the cloud computing platform, it is important to identify all of the enterprises that may touch the enterprise data and understand where they are located. This includes not only the CSP, but all vendors that are in the critical path of the CSP. Background checks on these enterprises are important to ensure that data are not being hosted by an organization that is incapable of responding to outages or providing business continuity or that is engaging in malicious or fraudulent activity.
4. **Culture, Ethics and Behaviour:**
   - *CSP business viability*—As cloud computing continues to mature, some CSPs will thrive and others will go out of business. Clients need to consider the risk and how data and applications can be easily transferred back to the traditional enterprise or to another CSP.
   - *Screening of other cloud computing clients*—By definition, CSPs leverage their cloud computing technology for many clients concurrently to maximize revenue. Clients should consider whether the other clients who share the same servers—and, in the case of SaaS, the same application and data files—are of the same repute as their own enterprises.
5. **Information:**
   - *Cloud data ownership*—Contract agreements may state that the CSP owns the data placed in the cloud computing environment that it maintains. The CSP may also require significant service fees for data to be returned to clients if and when a cloud computing services agreement is terminated.
   - *Record protection for forensic audits*—Clients must also consider the availability of data and records, if required for forensic audits. Since data may have been commingled and migrated among multiple servers located widely apart, it may be possible that the data for a specific point in time cannot be located. Furthermore, local authorities may impound a cloud computing server to assess court-warranted data records of a suspect client, taking with it the data of all the cloud computing clients sharing this impounded server.

- *Data disposal for current SaaS or PaaS applications*—When an application and data are transferred from one server to another, as would be expected with dynamic scalability, the earlier application and data files may remain and may not be erased. Their space on the original hard drives is now available for overwrites. The original data files may still be available for copying up to the third rewrite of the original disk space. This remaining copy of data may be useful in the case of an emergency; however, it presents customers with the dilemma of ensuring that confidential data are permanently destroyed in the event of a contract termination. Customers need to ensure that this confidentiality is implemented by including language in the contract that provides for immediate data erasure upon contract termination.

6. **Services, Infrastructure and Applications:**
   - *Data location*—Regardless of the deployment model selected, customers may not know the physical location of the server that is used to store and process their data and applications. Cloud computing technology allows cloud servers to reside anywhere. From a technology standpoint, location becomes mostly irrelevant. However, for many compliance and data governance requirements, the physical location of the cloud computing server hosting user data is a critical issue. While the data may reside anywhere, it is important to understand that many CSPs can also specifically define where data are to be located—down to the server, data center and country levels.
   - *Commingled data*—Depending on the deployment and service model, clients might use the same application on the same server concurrently. This may result in data from more than one client being stored in the same data files. SaaS providers might claim that each data field has an appropriate metatag affixed to keep clients' commingled data separate. Encryption is another control that can assist in data confidentiality; however, users need to ascertain the specifics of encryption key management and the process used to unencrypt data prior to being processed.
   - *Identity and access management (IAM)*—Current CSPs may not develop and implement adequate user access privilege controls. With ever more sophisticated applications going online and available for access by enterprise users, partners and clients, highly granular, least-privilege-based user access tools are required.
   - *Disaster recovery (DR)*—DR is a concern for potential cloud customers. In traditional hosting or colocation sites, customers know exactly where their data are located if they need to quickly retrieve them. The cloud model can change so that public CSPs may outsource capabilities to third parties who may also outsource—the original CSP may not be the CSP ultimately holding the data. Contracts should detail any testing or recovery time requirements.
7. **People, Skills and Competencies:**
   - *Lock in with CSP proprietary application program interfaces (APIs)*—Like the proprietary software vendor applications in the 1970s, many CSPs currently implement their applications using proprietary APIs. This makes transitioning between CSPs extremely difficult, time-consuming and labor-intensive. Uploading data into a cloud SaaS is easier and less costly than transferring data from one CSP with proprietary APIs to a replacement CSP.

The cloud computing client is responsible for:
- Assessing the cloud computing risk and controls, especially when using a public cloud computing delivery model
- Ensuring that the enterprise's sensitive data remains authentic, accurate and available
- Ensuring that the data that is applicable to regulations meets the specific compliance requirements

Cloud computing challenges to consider:[7]
- Data location
- Commingled data
- Security policy/procedure transparency
- Cloud data ownership
- Lock-in with CSP proprietary APIs
- Record protection for forensic audits
- Identity and access management (IAM)
- Screening of other cloud computing clients
- Compliance requirements
- Data disposal
- Portability
- CSP viability
- Backup and rollout capabilities

Figure 5 provides an overview of cloud risk scenarios. See appendix D. Cloud Risk Scenarios for further information.

Figure 5—Risk Scenario Overview

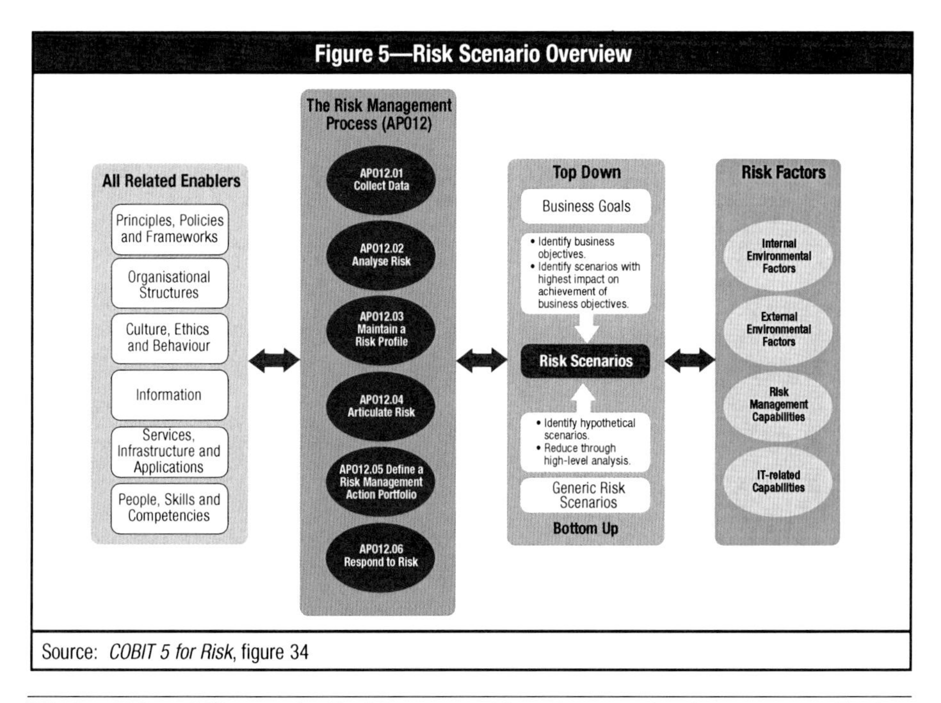

Source: *COBIT 5 for Risk*, figure 34

[7] ISACA, *IT Control Objectives for Cloud Computing* webinar, 2011, *www.slideshare.net/ramsesgallego/it-control-cloudwebinaraugust11th*

Risk Assessment When Migrating to the Cloud

The chief information security officer (CISO) or the information security manager (ISM) is responsible for being aware of the current risk affecting the assets of the enterprise and for understanding how the migration to the cloud will affect those assets and the current level of risk. In absence of a CISO or ISM, these are the responsibilities of a similar control organization/function within the enterprise.

The impact of a migration to the cloud depends on the cloud service model and deployment model being considered. The combination of service model and deployment model can help to identify an appropriate balance for organizational assets, e.g., choosing a private cloud deployment model can help balance the risk related to commingled data (multitenancy). In the previous section, the general cloud computing business challenges were enumerated. **Figure 6** lists risk-decreasing and risk-increasing factors by service model and deployment model. These risk factors are linked to actual threats and mitigating actions in chapter 4.

To facilitate a better understanding of the issues specific to the cloud, common risk factors (increasing or decreasing) that are not linked solely to cloud infrastructures, but apply to all types of infrastructure, are not covered in this publication. Examples of such risk factors include external hacking, malicious insiders, mobile computing vulnerabilities, virus and malicious code and business impact due to provider inability.

Risk Factors by Service Model[8]

Figure 6—Risk Factors by Service Model

Code	Risk Factor	Risk Event				Type	Description
		Availability	Loss	Theft	Disclosure		
IaaS							
S1.A	Scalability/elasticity	X				Decreasing	Lack of physical resources is no longer an issue. Due to the scalable nature of cloud technologies, the CSP can provide capacity on demand at low cost to support peak loads (expected or unexpected). Elasticity eliminates overprovisioning and underprovisioning of IT resources, allowing better cost optimization. This becomes a great advantage for resilience when defensive measures or resources need to be expanded quickly (e.g., during distributed denial-of-service [DDoS] attacks).
S1.B	Disaster recovery and backup	X	X			Decreasing	CSPs should already have in place, as common practice, disaster recovery and backup procedures. However, recovery point objective (RPO), recovery time objective (RTO), and backup testing frequency and procedures provided by the CSP should be consistent with the enterprise security policy.
S1.C	Patch management	X	X	X	X	Decreasing	Cloud infrastructures are based on hypervisors and are controlled through a central hypervisor manager or client. The hypervisor manager allows the necessary patches to be applied across the infrastructure in a short time, reducing the time available for a new vulnerability to be exploited.
S1.D	Legal transborder requirements				X	Increasing	CSPs are often transborder, and different countries have different legal requirements, especially concerning personal private information. The enterprise might be committing a violation of regulations in other countries when storing or transmitting data within the CSP's infrastructure without the necessary compliance controls. Furthermore, government entities in the hosting country may require access to the enterprise's information, with or without proper notification.

[8] ISACA, *Security Considerations for Cloud Computing*, USA, 2012, pp. 17-24, *www.isaca.org/Knowledge-Center/Research/ResearchDeliverables/Pages/Security-Considerations-for-Cloud-Computing.aspx*

Figure 6—Risk Factors by Service Model *(cont.)*

Code	Risk Factor	Risk Event				Type	Description
		Availability	Loss	Theft	Disclosure		
IaaS *(cont.)*							
S1.E	Multitenancy and isolation failure			X	X	Increasing	One of the primary benefits of the cloud is the ability to perform dynamic allocation of physical resources when required. The most common approach is a multitenant environment (public cloud), where different entities share a pool of resources, including storage, hardware and network components. All resources allocated to a particular tenant should be "isolated" and protected to avoid disclosure of information to other tenants. For example, when allocated storage is no longer needed by a client, it can be freely reallocated to another enterprise. In that case, sensitive data could be disclosed if the storage has not been scrubbed thoroughly (e.g., using forensic software).
S1.F	Lack of visibility surrounding technical security measures in place	X	X	X	X	Increasing	For any infrastructure, intrusion detection system (IDS)/intrusion prevention system (IPS) and security incident and event management (SIEM) capabilities must be in place. It is the responsibility of the CSP to provide these capabilities to its customers. To ensure that there are no security gaps, the security policy and governance of the CSP should match those of the enterprise.
S1.G	Absence of DRP and backup	X	X			Increasing	The absence of a proper disaster recovery plan (DRP) or backup procedures implies a high risk for any enterprise. CSPs should provide such basic preventive measures aligned with the enterprise's business needs (in terms of RTO/RPO).
S1.H	Physical security			X	X	Increasing	In an IaaS model, physical compute resources are shared with other entities in the cloud. If physical access to the CSP's infrastructure is granted to one entity, that entity could potentially access information assets of other entities. The CSP is responsible for applying physical security measures to protect assets against destruction or unauthorized access.
S1.I	Data disposal				X	Increasing	Proper disposal of data is imperative to prevent unauthorized disclosure. If appropriate measures are not taken by the CSP, information assets could be sent (without approval) to countries where the data can be legally disclosed due to different regulations concerning sensitive data. Disks could be replaced, recycled or upgraded without proper cleaning so that the information still remains within storage and can later be retrieved. When a contract expires, CSPs should ensure the safe disposal or destruction of any previous backups.

Figure 6—Risk Factors by Service Model *(cont.)*

Code	Risk Factor	Risk Event				Type	Description
		Availability	Loss	Theft	Disclosure		
IaaS *(cont.)*							
S1.J	Offshoring infrastructure	X	X	X	X	Increasing	The move to offshoring of key infrastructure expands the attack surface area considerably. In practice, the information assets in the cloud still need to integrate back to other noncloud-based assets within the boundaries of the organization. These communications (normally done through border gateway devices) could be insecure, exposing both the cloud and internal infrastructures.
S1.K	VM security maintenance	X	X	X	X	Increasing	IaaS providers allow consumers to create virtual machines (VMs) in various states (e.g., active, running, suspended and off). Although the CSP could be involved, the maintenance of security updates is generally the responsibility of the customer only. An inactive VM could be easily overlooked and important security patches could be left unapplied. This out-of-date VM could become compromised when activated.
S1.L	Cloud provider authenticity	X	X	X	X	Increasing	Although communications between the enterprise and the cloud provider can be secured with technical means (encryption, VPN, mutual authentication, etc.) it is a consumer's responsibility to check the identity of the cloud provider to ensure that is not an imposter.
PaaS							
S2.A	Short development time	X	X			Decreasing	Using the SOA library provided by the CSP, applications can be developed and tested within a reduced timeframe because SOA provides a common framework for application development.
S2.B	Application mapping			X	X	Increasing	If current applications are not perfectly aligned with the capabilities provided by the CSP, additional undesirable features (and vulnerabilities) could be introduced.
S2.C	SOA-related vulnerabilities	X	X	X	X	Increasing	Security for SOA presents new challenges because vulnerabilities arise not only from the individual elements, but also from their mutual interaction. Because the SOA libraries are under the responsibility of the CSP and are not completely visible to the enterprise, there may exist unnoticed application vulnerabilities.
S2.D	Application Disposal			X	X	Increasing	When applications are developed in a PaaS environment, originals and backups should always be available. In the event of contract termination, the details of the application could be disclosed and used to create more selective attacks on applications.

Figure 6—Risk Factors by Service Model *(cont.)*

Code	Risk Factor	Risk Event				Type	Description
		Availability	Loss	Theft	Disclosure		
SaaS							
S3.A	Improved security	X	X			Decreasing	CSPs depend on the good reputation of their software capabilities to maintain their SaaS offering. Consequently, they introduce additional features to improve the resilience of their software (e.g., security testing or strict versioning) or to inform users about the exact state of their business application (e.g., specific software logging and monitoring).
S3.B	Application patch management	X	X			Decreasing	Due to the fact that the SaaS application service is managed globally and only by the CSPs, application patch management is more effective, allowing patches to be deployed in little time with limited impact.
S3.C	Data ownership	X	X		X	Increasing	The CSP provides the applications and the customer provides the data. If data ownership is not clearly defined, the CSP could refuse access to data when required or even demand fees to return the data once the service contracts are terminated.
S3.D	Data disposal			X	X	Increasing	In the event of a contract termination, the data fed into the CSP's application must be erased immediately using forensic tools to avoid disclosures and confidentiality breaches.
S3.E	Lack of visibility into software SDLC	X	X	X	X	Increasing	Enterprises that use cloud applications have little visibility into the software system development life cycle (SDLC). Customers do not know in detail how the applications were developed and what security considerations were taken into account during the SDLC. This could lead to an imbalance between the security provided by the application and the security required by customers/users.
S3.F	IAM		X	X	X	Increasing	To maximize their revenues, CSPs offer their services and applications to several customers concurrently. Those customers share servers, applications and, eventually, data. If data access is not properly managed by the CSP application, one customer could obtain access to another customer's data.

Figure 6—Risk Factors by Service Model *(cont.)*

Code	Risk Factor	Risk Event				Type	Description
		Availability	Loss	Theft	Disclosure		
SaaS *(cont.)*							
S3.G	Exit strategy	X	X			Increasing	Currently, there is very little available in terms of tools, procedures or other offerings to facilitate data or service portability from CSP to CSP. This can make it very difficult for the enterprise to migrate from one CSP to another or to bring services back in-house. It can also result in serious business disruption or failure should the CSP go bankrupt, face legal action, or be the potential target for an acquisition (with the likelihood of sudden changes in CSP policies and any agreements in place). If the customer-CSP relationship goes sour and the enterprise wants to bring the data back in-house, the question of how to securely render the data becomes critical because the in-house applications have been decommissioned or "sunsetted" and there is no application available to render the data.
S3.H	Broad exposure of applications	X	X		X	Increasing	In a cloud environment, the applications offered by the CSP have broader exposure, which increases the attack space; additionally, it is quite common that those applications still need to integrate back to other noncloud applications within the boundaries of the enterprise. Standard network firewalls and access controls are sometimes insufficient to protect the applications and their external interactions. Additional security measures are required.
S3.I	Ease to contract SaaS	X	X	X	X	Increasing	Business organizations may contract cloud applications without proper procurement and approval oversight, thus bypassing compliance with internal enterprise policies.
S3.J	Lack of control of the release management process	X	X			Increasing	As described before, CSPs are able to introduce patches in their applications quickly. These deployments are often done without the approval (or even the knowledge) of the application users for merely practical reasons: If an application is used by hundreds of different enterprises, it would take an extremely long time for a CSP to look for the formal approval of every customer. In this case, the enterprise could have no control (or no view) of the release management process and could be subject to unexpected side effects.
S3.K	Browser vulnerabilities			X	X	Increasing	As a common practice, applications offered by SaaS providers are accessible to customers via secure communication through a web browser. Web browsers are a common target for malware and attacks. If the customer's browser becomes infected, the access to the application can be compromised as well.

Risk Factors by Deployment Model

Cloud deployment models do not have the same abstraction as cloud service models. That is, risk is not cumulative, but particular to each model. However, "trust" among the different entities (CSP, customers, CSP third-party service providers, etc.) is an important factor—not just trust between the CSP and the customer, but enough trust in the other tenants sharing computing resources hosting the enterprise's information assets. If a user abuses the infrastructure and services of the public cloud, the entire infrastructure might be at risk of failure, theft or seizure (for forensics), including the services used by other enterprises. It is important as part of the decision process to carefully consider which assets can securely be hosted in a public cloud and which cannot.

Figure 7 lists the risk factors by deployment model.

Figure 7—Risk Factors by Deployment Model

Code	Risk Factor	Risk Event				Type	Description
		Availability	Loss	Theft	Disclosure		
Public Cloud							
D1.A	Public reputation	X	X	X	X	Decreasing	Providers of public cloud services are aware that they are generally perceived as more "risky." It is critical for them to ensure a good reputation because a secure provider or customers will simply move elsewhere.
D1.B	Full sharing of the cloud (data pooling)	X	X	X	X	Increasing	The cloud infrastructure is shared by multiple tenants of the CSP. These tenants have no relation to the enterprise or other tenants in the same space, therefore no common interests and concern for security.
D1.C	Collateral damage	X	X	X	X	Increasing	If one tenant of a public cloud is attacked, there could be an impact to the other tenants of the same CSP, even if they are not the intended target (e.g., DDoS). Another possible scenario of collateral damage could be a public cloud IaaS that is affected by an attack, exploiting vulnerabilities of software installed by one of the tenants.

Figure 7—Risk Factors by Deployment Model *(cont.)*

Code	Risk Factor	Risk Event				Type	Description
		Availability	Loss	Theft	Disclosure		
Community Cloud							
D2.A	Same group of entities	X	X	X	X	Decreasing	The component of "trust" among the entities in a community cloud makes the level of risk lower than in a public cloud. (However, it remains higher than in a private cloud.)
D2.B	Dedicated access for the community			X	X	Decreasing	Dedicated access can be configured for authorized community users only.
D2.C	Sharing of the cloud		X	X	X	Increasing	Different entities may have different security measures or security requirements in place, even if they belong to the same enterprise. This could render an entity at risk because of the faulty procedures or SLAs of another entity, or simply because of differing security levels for the same type of data.
Private Cloud							
D3.A	Can be built on premises	X	X	X	X	Decreasing	Physical or location-related considerations can be more closely controlled by the enterprise because the cloud infrastructure can be located on the enterprise's premises. Global enterprise security policies would apply.
D3.B	Performance	X	X			Decreasing	Affects onsite private clouds. Because the private cloud is deployed inside the firewall on the enterprise's intranet, transfer rates are dramatically increased (fewer nodes to cross). Storage capacity can also be higher; private clouds usually start with a few terabytes and can be increased by adding disks.
D3.C	Application compatibility	X	X			Increasing	While applications that have already been confirmed to be virtualization-friendly are likely to run well in a private cloud environment, problems can occur with older and/or customized software that assumes direct access to resources. Larger applications that currently run on dedicated specialized clusters with hardwiring into proprietary runtime and management environments may also be questionable candidates for migration, at least until standards settle and vendors take steps to make their solutions private-cloud-compatible. In the meantime, compatibility testing and remediation are critical.

Figure 7—Risk Factors by Deployment Model *(cont.)*

Code	Risk Factor	Risk Event				Type	Description
		Availability	Loss	Theft	Disclosure		
Private Cloud *(cont.)*							
D3.D	Investments required	(can be triggered by cost)				Increasing	Making a business case for shared infrastructure and the necessary training or recruitment to acquire associated skills is notoriously hard at the best of times. Although the word "cloud" has a high profile, messages from vendors and service providers are often confusing and contradictory, making seeking support from senior stakeholders even more of an issue. If the head of finance thinks cloud is all about getting rid of infrastructure, it can be difficult to explain that investments are needed in new equipment, software, and tools. The enterprise must conduct a business case and cost-benefit analysis to determine whether the cloud is a viable solution to meet specific business requirements, and justify any expenses.
D3.E	Cloud IT skills required	(can be triggered by cost)				Increasing	Affects onsite private clouds. Building a private cloud within the enterprise infrastructure seems the best option in terms of security. However, the maintenance of cloud infrastructures requires specific cloud IT skills in addition to the traditional IT skills, thus increasing the required initial investments.
Hybrid Cloud							
D4.A	Cloud interdependency	X	X	X	X	Increasing	If the enterprise mixes two or more different types of clouds, strict identity controls and strong credentials will be needed to allow one cloud to have access to another. This is similar to a common network infrastructure problem: how to allow access from a low-level security zone to a high-level security zone.

3. Governance and Management in the Cloud

Cloud computing is a combination of technologies through which dynamically scalable and often virtualized resources are provided as a service over the Internet. Users do not need to have knowledge of, expertise in, or control over the infrastructure in the cloud. The governance in the cloud needs to be adapted accordingly to ensure continuing performance and compliance against agreed-on enterprise direction, goals and objectives.

Therefore, the role of the chief information officer (CIO) is transitioning from managing operations to managing IT as a service value chain. Previously, the CIO may have run the internal "IT factory;" now, both enterprises and external service providers have become producers and consumers of services. With cloud computing, the CIO must weave together and optimize this value chain to best support various customers and enable an enterprise's business. The cloud is accelerating and mandating the transition and, therefore, governance is mandatory.[9]

Governance and Management of Enterprise IT (GEIT)

When enterprises decide to use cloud services for some or all IT services, business processes are impacted, which makes governance more critical than ever. The following are just a few reasons why enterprises should implement and maintain sound governance:[10]

- Manage increasing risk effectively, including security, compliance, privacy, projects and partners.
- Ensure continuity of critical business processes that now extend beyond the data center.
- Communicate clear enterprise objectives internally and to third parties.
- Adapt easily. Flexibility, scalability and services are changed in the cloud, enabling the enterprise and business practices to adjust, to create new opportunities and reduce costs.
- Facilitate continuity of IT knowledge, which is essential to sustain and grow the business.
- Handle a myriad of regulations.

For enterprises to gain benefit from the use of cloud computing, a clear governance strategy and management plan must be developed. The governance strategy should set the direction and objectives for cloud computing within the enterprise, and the management plan should execute the achievement of the objectives. In support of the establishment of a governance strategy, appendix F contains a cloud ERM governance checklist. The checklist provides cloud-related questions that a board of directors should consider asking in its governance oversight role.

---

[9] Stroud, Robert; "Providing Governance in a Rapidly Changing World," ISACA Euro Computer Audit, Control and Security (EuroCACS) conference 2010, Budapest, Hungary

[10] *Ibid.*

Business consultants have long recognized that the various functional departments of highly successful enterprises tend to interact nearly seamlessly. The key to this successful departmental integration is recognizing and leveraging the interrelationships and interdependencies among departments. This originates from the tone at the top and is manifested and reinforced via cross-departmental corporate governance programs.

IT, with its particular requirements and terminology, has historically been viewed as a cost center rather than a corporate asset. However, the cloud presents the opportunity to fully align IT with the goals and objectives of the enterprise as a whole. To "sit at the table" of the enterprise's governance programs, IT must adapt to methodologies and the language used in other areas of the enterprise's governance.

ISACA's definition of governance: Governance ensures that stakeholder needs, conditions and options are evaluated to determine balanced, agreed-on enterprise objectives to be achieved; setting direction through prioritisation and decision making; and monitoring performance and compliance against agreed-on direction and objectives.[11] GEIT is primarily concerned with the evaluation, direction and monitoring of benefits delivery to the business, resource optimization and the management of IT-related risk. **Figure 8** illustrates risk duality in GEIT.

Figure 8—Risk Duality

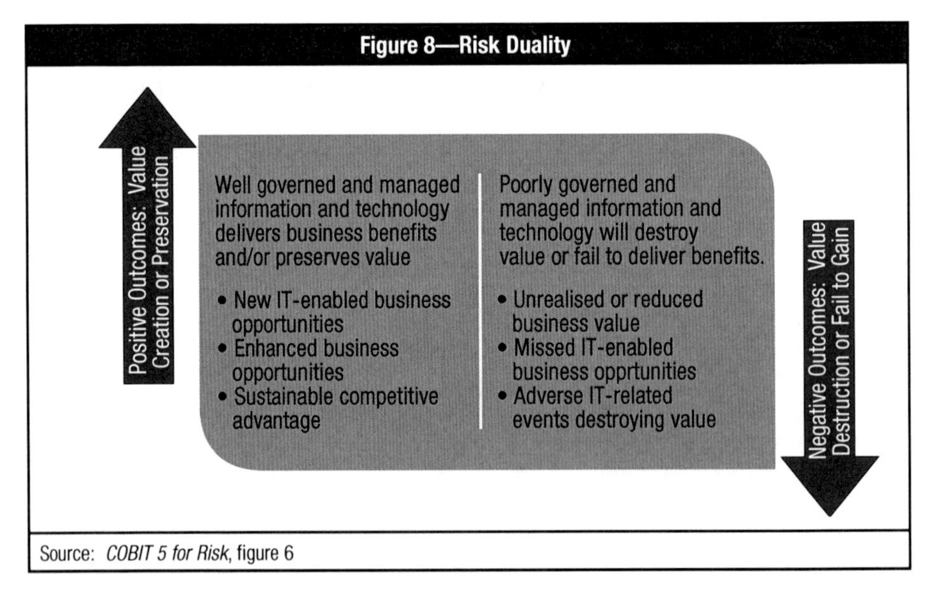

Source: *COBIT 5 for Risk*, figure 6

Governance From the Top[12]

To establish a clear direction that is aligned with enterprise strategy, members of the board of directors need to have a clear understanding of cloud computing benefits and how to maximize them through effective end-to-end governance practices.

[11] ISACA, COBIT 5, USA, 2012, *www.isaca.org/cobit*

[12] ISACA, "Cloud Governance: Questions Boards of Directors Need to Ask," USA, 2013, *www.isaca.org/Knowledge-Center/Research/ResearchDeliverables/Pages/Cloud-Governance-Questions-Boards-of-Directors-Need-to-Ask.aspx*

This requires the board to see cloud computing not as an IT project, but rather as a business technology strategy. This understanding helps to ensure that stakeholder needs are considered and met while risk and resource utilization are optimized.

The following questions help to identify the strategic value that cloud services may provide to the enterprise and the impact that cloud could have on enterprise resources and controls. Upper management (CEO, CFO, CIO or CTO) must be prepared to answer these questions in terms that can be easily related to the business benefits cloud computing will provide.

1. **Do management teams have a plan for cloud computing? Have they weighed value and opportunity costs?** The risk of cloud adoption may be inconsequential when compared to the lost opportunity to transform the enterprise with effective and strategic use of cloud computing. The loss can be particularly great when competitive enterprises take steps to leverage those same opportunities. From a strategic perspective, cloud computing can be a vehicle to:
   - **Gain competitive advantage.**
   - **Reach new markets.**
   - **Improve existing products and services.**
   - **Retain existing customers.**
   - **Increase productivity.**
   - **Contain cost.**
   - **Develop products or services that could not be possible without cloud services.**
   - **Break geographic barriers.**

2. **How do current cloud plans support the enterprise's mission?** Cloud services should support efforts to achieve business objectives, which are derived from stakeholder needs (as vetted by the leadership team). Cloud initiatives should have a clear and traceable link to the enterprise strategy so that the value expected from cloud services is clearly defined, accepted and measurable. This link also helps to determine the priority assigned to cloud initiatives and supports the development of metrics to measure results against expectations.

   Alignment between cloud objectives and enterprise objectives is critical for effective risk management and cost containment. The potential benefits of cloud services can be enticing, but with reward comes risk. The enterprise must decide whether the potential risk is within acceptable limits.

3. **Have executive teams systematically evaluated organizational readiness?** Pressure points result when:
   - **Cloud computing implementations conflict with enterprise culture.**
   - **Skills that are required to support cloud solutions are not available.**
   - **Cloud-related processes conflict with other established processes.**
   - **Organizational structure does not maximize cloud effectiveness or efficiency.**

Evaluating the readiness of the enterprise in anticipation of the adoption of cloud services avoids the need for after-the-fact culture, skill or process changes to remove unanticipated pressure points. A systematic readiness assessment can help management identify additional cost and risk that should be factored into the decision process. This readiness assessment should include the following:

- **Policies and procedures**—New policies and procedures that guide the adoption, management and proper use of cloud computing may be needed.
- **Processes**—Existing processes using traditional IT services may need to be reengineered to incorporate new activities that are related to using cloud services.
- **Organizational structures**—Cloud management may require new organizational capabilities or modifications to existing organizational structures, particularly in IT operations and support.
- **Culture and behavior**—Organizational culture and behavior can be critical for the successful adoption of cloud solutions.
- **Skills and competencies**—Procurement, legal, compliance and audit are some examples of functions that may need to develop necessary skills to manage cloud services from evaluation and sourcing to operations and retirement.

4. **Have management teams considered what existing investments might be lost in their cloud planning?** Cloud computing may not be an immediate and clean fit with the existing technology portfolio of the enterprise. The adoption of a cloud service may, for example, obviate already-made technology investments that have not reached their planned end date. The decision about when and how to realize that loss must be considered carefully. Areas to consider include:
   - **Processes**—The IT organization may need to adapt processes such as sourcing and change management.
   - **Culture and behavior**—Cloud services may demand faster turnaround from the IT organization, which may necessitate changes in internal processes and tools.
   - **Services, infrastructures and applications**—The enterprise may need to update data centers, software applications and network infrastructures, which may result in some level of lost investment being realized.
   - **Skills and competencies**—The IT organization will need to either develop or acquire the skills required to support users of cloud services, if those skills do not already exist within current staffing.
5. **Do management teams have strategies to measure and track the value of cloud return vs. risk?** Before deciding to adopt cloud computing, the board should give management teams the task of ensuring that proper reporting mechanisms are in place to measure value and risk aligned with enterprise goals.

Appendix F provides a governance checklist that can be used to capture these and other questions that should be considered an important part of decision process to adopt cloud computing and later the support the cloud management efforts.

ISACA's Governance and Management Framework

ISACA has developed several GEIT tools to assist IT executives and managers with integrating and aligning IT operations into the primary business focus of the enterprise. COBIT 5 (the framework) is the comprehensive GEIT framework of

ISACA that addresses every aspect of IT and integrates all of the main global IT standards. COBIT 5 starts from the stakeholder needs that drive the governance objective: value creation. COBIT 5 integrates the older Val IT, Risk IT and COBIT 4.1 into one comprehensive document. COBIT 5 provides continued real-time quantification of current IT investments, verifying if and why a specific IT investment has been successful. COBIT 5 is also used to assess the value of new or in-process development IT projects, monitoring the development and the operational costs.

The COBIT 5 Evaluate, Direct and Monitor (EDM) domain includes the following processes:

- EDM01 Ensure governance framework setting and maintenance.
- EDM02 Ensure benefits delivery.
- EDM03 Ensure risk optimization.
- EDM04 Ensure resource optimization.
- EDM05 Ensure stakeholder transparency.

Leveraging and Integrating the COBIT Family of Products

It is clear that there are strong links among COBIT 5 publications. *COBIT 5 for Risk, COBIT 5 for Information Security* and *COBIT 5 for Assurance* complement and extend the COBIT 5 guidance in the focus areas of risk management, information security and assurance. If these areas are the focus of the enterprise's GEIT implementation efforts, then *COBIT 5 for Risk* and *COBIT 5 for Information Security* can help to identify more specifically what should be addressed to enable better governance of risk management and information security management.

Brief descriptions of these publications are as follows:

- ***COBIT 5 for Information Security***—Provides guidance and principles for managing information security from a business-oriented approach. By looking at areas such as enterprise culture and processes, this publication helps security professionals to identify threats and risk that may not have been previously recognized until they were realized. In the cloud, *COBIT 5 for Information Security* enables IT professionals to look at additional threats that arise from third parties, such as CSPs, and understand the impact that those threats have on the rest of the business.
- ***COBIT 5 for Risk***—Provides guidance and principles for the risk function in an enterprise and builds on the COBIT 5 framework to help enterprises effectively identify, analyze, respond to and report on IT risk. Furthermore, this publication helps to reduce regulatory concerns, which allows more IT innovation to support new or expanded business initiatives. The implementation of *COBIT 5 for Risk* can also result in fewer operational surprises and failures, improved information quality, and increased stakeholder confidence.
- ***COBIT 5 for Assurance***—Provides insights on how to use the different COBIT 5 components and related concepts for planning, scoping, executing and reporting on various types of IT assurance initiatives, among which are cloud computing assurance initiatives. Furthermore, this publication helps to obtain a view of the extent to which the value objective of the enterprise for cloud computing—delivering benefits while optimizing risk and resource use—is achieved.

While these tools have not been designed specifically for cloud environments, the principles are applicable. In fact, cloud computing may amplify the importance these tools have in an enterprise, because the impact that the cloud has on the business may change business processes. In traditional IT environments, everyone in the business must go to the IT department to obtain IT-related services; with the capabilities that the cloud provides enterprises, employees can go directly to a CSP to acquire services. This elevates the enterprise's level of risk. Having tools such as those mentioned from ISACA will help the enterprise to implement repeatable processes and appropriate control levels.

Because *COBIT 5 for Risk* and *COBIT 5 for Information Security* extend and complement all seven enablers in COBIT 5 with enablers that manage risk and information security, including better business involvement, all three can be used together to help create a set of end-to-end IT-related process practices and other enablers. This will help to integrate all business and IT activities for effective GEIT. *COBIT 5 for Assurance* can be used afterward to assess whether the expected business value and benefits of the cloud initiatives are achieved.

Together, COBIT 5, *COBIT 5 for Assurance, COBIT 5 for Risk* and *COBIT 5 for Information Security* provide an effective way to understand business and governance priorities and requirements; this knowledge can then be used when implementing cloud services. This approach also enhances the preparation of business cases for governance improvements, obtaining the support of stakeholders, and realizing and monitoring the expected benefits.[13] This approach, from understanding business and governance priorities to realizing and monitoring the benefits, can be summarized in the top-down flow shown in **figure 9**. The COBIT 5 family drives the actions to take, which are supported by a cascade of the stakeholder needs to enterprise goals, to IT-related goals and, finally, to enabler goals.

There are many stakeholders interested in GEIT who need to collaborate to achieve the overall business goal of improved IT performance (see also COBIT 5 process EDM05). When moving to a cloud environment, this group of stakeholders may expand because it may affect processes within other business units and IT. If the enterprise does not already have a robust GEIT program in place, the move toward a cloud solution provides a good opportunity to develop one. If there is a mature program in place, it will need to be reviewed, assessed, and most likely adjusted to account for changes in business processes, security and risk. When creating or adjusting a GEIT program, key success factors for implementation are:

- Top management should provide the direction and mandate for GEIT.
- All parties supporting the governance process should understand the business and IT objectives.
- The GEIT program should ensure that there is effective communication and enablement of the necessary organizational and process changes.
- The enterprise should tailor the GEIT program and good practices to fit the enterprise purpose and design.
- The GEIT program should focus on quick wins and the prioritization of the most beneficial improvements that are easiest to implement.

[13] ISACA, *COBIT 5 Implementation*, USA, 2012, *www.isaca.org/cobit*

Figure 9—COBIT 5 Goals Cascade Overview

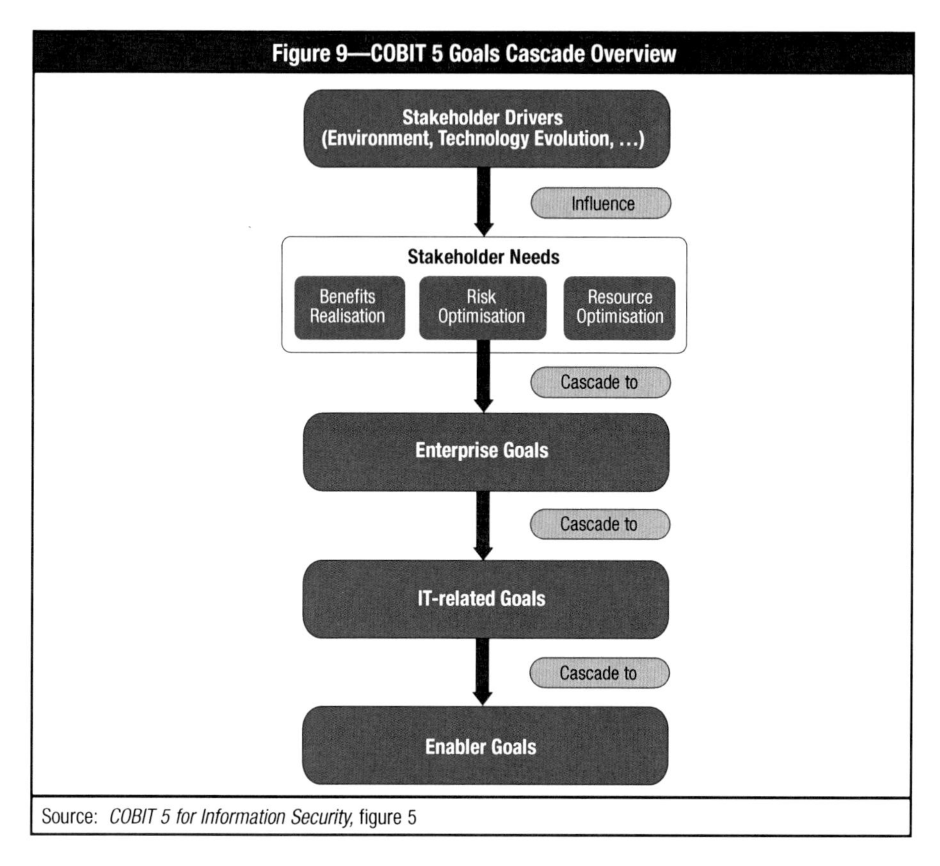

Source: *COBIT 5 for Information Security*, figure 5

Cloud Governance Advantages

Applying cloud governance as described in the COBIT 5 goals cascade (**figure 9**) allows the enterprise's cloud computing initiative to provide value by:

- Realizing the benefits envisioned at the start of the cloud computing initiative
- Managing the risk an enterprise is willing to take during the cloud computing initiative
- Ensuring an optimal use and management of the variety resources involved in a cloud computing initiative

Benefits Realization

Benefits enablement is a key consideration of a cloud program. The cloud can provide enterprises with many business benefits by enabling new business initiatives or more efficient operations.

New cloud initiatives may include:

- A broad-based consumer-oriented online marketing campaign that requires highly elastic processing power and provides video feeds to consumers on demand
- A one-time information processing exercise to catalog current stored data for easy query selections
- An informational web site that provides area residents with up-to-date emergency guidance in the event of a local emergency, disaster or impending weather event

More efficient operations involving the cloud may include:
- A transfer from an internal to a cloud-based CRM program that allows the enterprise's sales force to utilize and collaborate on a best-of-breed application for less cost than the current in-house application
- An upgrade to an industry-recognized best-practice online enterprise resource planning (ERP) application to coordinate product scheduling plans with suppliers and resellers
- A consumer-facing e-commerce web page with near-infinite scalability for market volume recognized to be widely variable over the course of a day
- A marketing/promotional blog that is prepared for variable scalability, but also includes smart profiling of registered client bloggers that emails them when a message in their interest space has been posted

Risk Optimization

IT risk in the cloud needs to be managed on a variety of fronts, distinguished here via the following three IT risk categories that are shown in **figure 10**:
1. **IT benefit/value enablement risk**—Associated with missed opportunities to use technology to improve efficiency or effectiveness of business processes or as an enabler for new business initiatives
2. **IT programme and project delivery risk**—Associated with the contribution of IT to new or improved business solutions, usually in the form of projects and programmes as part of investment portfolios
3. **IT operations and service delivery risk**—Associated with all aspects of the business as usual performance of IT systems and services, which can bring destruction or reduction of value to the enterprise

Figure 10—IT Risk Categories

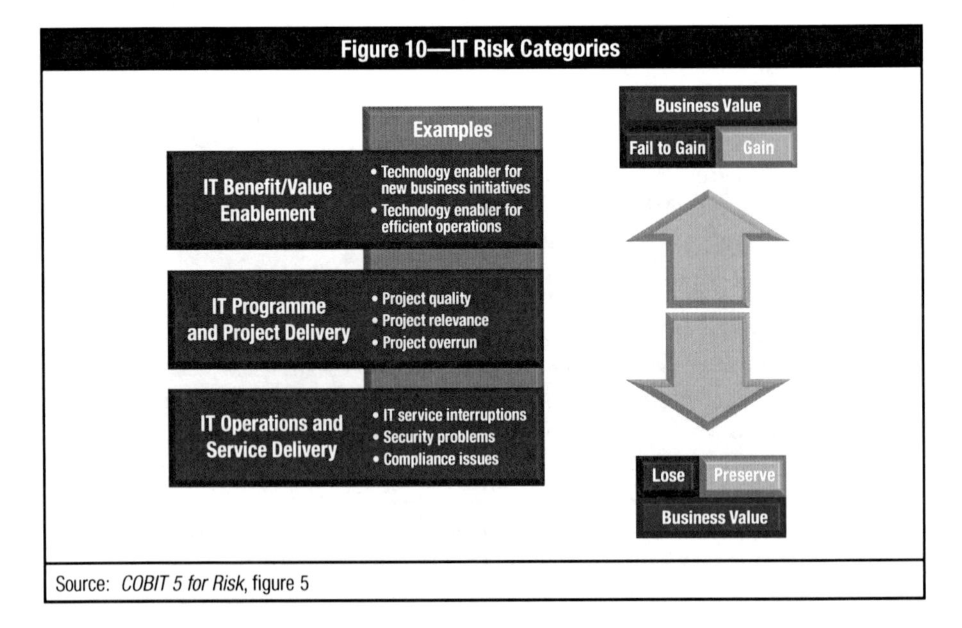

Source: *COBIT 5 for Risk*, figure 5

Figure 10 also shows that for all categories of downside IT risk ("Fail to Gain" and "Lose" business value) there is an equivalent upside ("Gain" and "Preserve" business). For example:

- **IT benefit/value enablement risk:**
  - Will the cloud program provide a user environment that matches or exceeds the experience offered by competitors? Will current and new users welcome and embrace the experience?
  - Will the cloud program be completed and then maintained within estimated budgets, providing or exceeding the expected value?
- **IT programme and project delivery risk:**
  - Can the enterprise's IT organization successfully execute the planned new cloud program successfully, on budget and on schedule? Will the cloud program provide the expected business benefit? Would not moving to the cloud cause additional levels of risk?
  - Will the cloud project provide a user environment—for internal, partner and customer users—that is well thought out, intuitive and welcoming to use?
  - Will the environment address the prespecified business needs and be able to accommodate requirements identified post implementation?
- **IT operations and service delivery risk:**
  - Will the operational support provided by the CSP maintain expected performance?
  - Can the new cloud program meet current and future regulatory requirements?

COBIT 5 for Risk defines a structured way to identify, analyze and respond to risk via risk scenarios. Based on the three aggregated IT risk categories described previously, specific and detailed risk scenarios for a possible cloud initiative can be identified, as shown in appendix D.

Resource Optimization

Cloud computing initiatives tend to impact a wider variety of people inside (and sometimes even outside) the enterprise, compared to other services being outsourced. Most enterprises will outsource services that are not core to them, thus impacting a smaller amount of people. With cloud computing, however, most of the time enterprises target their core services and take them to the cloud to enjoy the benefits of cloud computing for their core businesses. Obviously this will impact a great number of people throughout the entire enterprise, hence, requiring adequate preparation and management of all resources to keep ensuring operability.

Using COBIT 5 to Manage the Cloud

The use of proven frameworks can help to enable an enterprise to realize expected benefits from the cloud. ISACA's COBIT 5 has evolved to be a well-recognized IT governance and management framework. Extending its use to cloud governance is a logical step because COBIT 5 is flexible and allows for innovation. The COBIT 5 framework:

- Is platform-agnostic, both in type and complexity
- Has sufficient depth to address nearly all the technical aspects of cloud computing

- Provides clearly defined process activity measures—not just binary ("present" or "not present"), but rather a capability model scale of 0 through 5

COBIT 5 addresses governance and management of enterprise IT throughout an entire program life cycle. COBIT 5 ensures systematic governance and management through the use of seven interconnected enablers. Enablers are factors that, individually and collectively, influence whether something will work—in this case, governance and management over cloud computing. Enablers are driven by the goals cascade, i.e., higher-level IT-related goals define what the different enablers should achieve.

As shown in **figure 11**, the COBIT 5 framework describes seven categories of enablers:

1. **Principles, Policies and Frameworks** are the vehicle to translate the desired behavior into practical guidance for day-to-day management.
2. **Processes** describe an organized set of practices and activities to achieve certain objectives and produce a set of outputs in support of achieving overall IT-related goals.
3. **Organisational Structures** are the key decision-making entities in an enterprise.
4. **Culture, Ethics and Behaviour** of individuals and of the enterprise are very often underestimated as a success factor in governance and management activities.
5. **Information** is pervasive throughout any organization and includes all information produced and used by the enterprise. Information is required for keeping the organization running and well governed, but at the operational level, information is very often the key product of the enterprise itself.
6. **Services, Infrastructure and Applications** include the infrastructure, technology and applications that provide the enterprise with information technology processing and services.
7. **People, Skills and Competencies** are linked to people and are required for successful completion of all activities and for making correct decisions and taking corrective actions.

Figure 11—COBIT 5 Enterprise Enablers

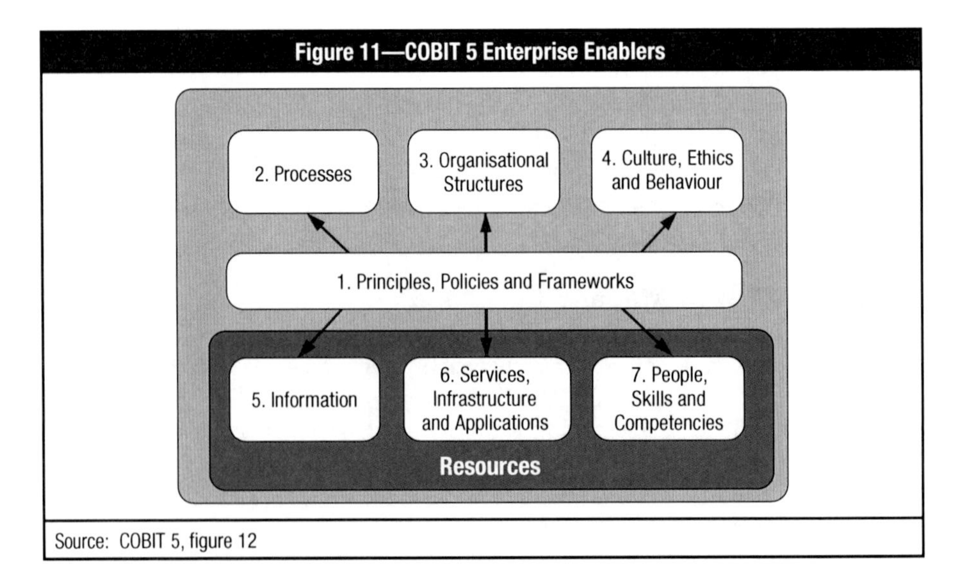

Source: COBIT 5, figure 12

A COBIT-based assessment provides the enterprise with a set of effective measures for gauging and controlling activities that are related to a cloud deployment or utilization. This assessment can be based on the set of four common enabler dimensions, which are shown in **figure 12** and described in the following paragraphs:

- **Stakeholders**—Each enabler has stakeholders (parties who play an active role and/or have an interest in the enabler). For example, processes have different parties who execute process activities and/or who have an interest in the process outcomes; organizational structures have stakeholders, each with his/her own roles and interests that are part of the structures. Stakeholders can be internal or external to the enterprise, having their own, sometimes conflicting, interests and needs. Stakeholder needs translate the stakeholder's business model into enterprise goals, which in turn translate to IT-related goals for the enterprise.
- **Goals**—Each enabler has a number of goals, and enablers provide value by the achievement of these goals. Goals can be defined in terms of:
  - Expected outcomes of the enabler
  - Application or operation of the enabler itself

  The enabler goals are the final step in the COBIT 5 goals cascade. Goals can be further split up in different categories:
  - **Intrinsic quality**—The extent to which enablers work accurately, objectively and provide accurate, objective and reputable results
  - **Contextual quality**—The extent to which enablers and their outcomes are fit for purpose given the context in which they operate. For example, outcomes should be relevant, complete, current, appropriate, consistent, understandable and easy to use.
  - **Accessibility and security**—The extent to which enablers and their outcomes are accessible and secured, such as:
    - Enablers are available when, and if, needed.
    - Outcomes are secured, i.e., access is restricted to those entitled and needing it.
- **Life cycle**—Each enabler has a life cycle, from inception through an operational/useful life until disposal. This applies to information, structures, processes, policies, etc. The phases of the life cycle consist of:
  - Plan (includes concepts development and concepts selection)
  - Design
  - Build/acquire/create/implement
  - Use/operate
  - Evaluate/monitor
  - Update/dispose
- **Good practices**—For each of the enablers, good practices can be defined. Good practices support the achievement of the enabler goals. Good practices provide examples or suggestions on how best to implement the enabler and the work products or inputs and outputs that are required. COBIT 5 provides examples of good practices for some enablers that are provided by COBIT 5, e.g., processes. For other enablers, guidance from other standards, frameworks, etc., can be used.

Figure 12—COBIT 5 Enablers: Generic

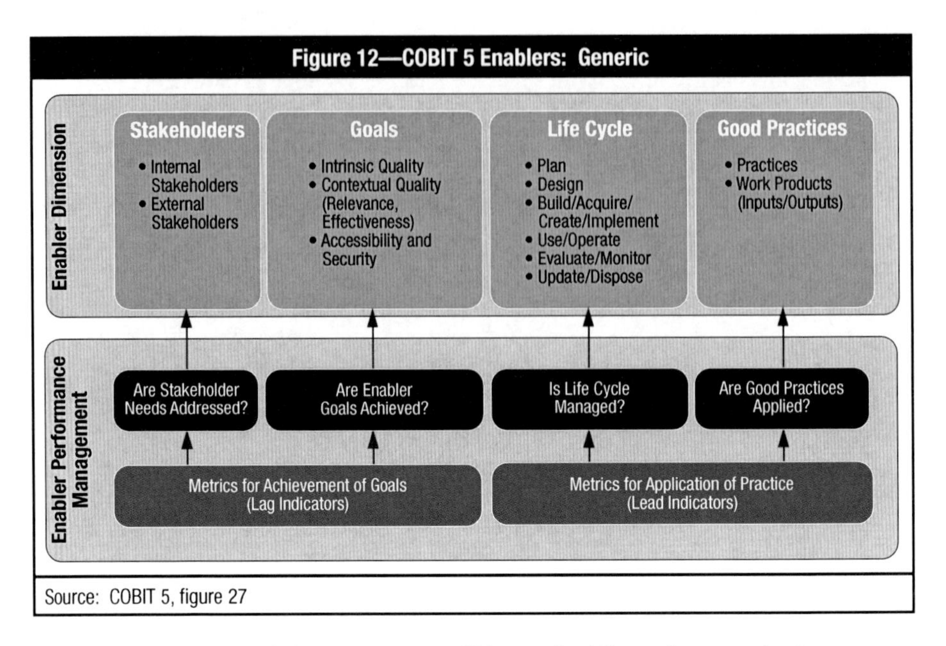

Source: COBIT 5, figure 27

From a historic point of view, COBIT 5 still has a significant focus on the Processes enabler. The COBIT 5 process reference model divides the governance and management processes of enterprise IT into two main process domains:

1. **Governance**—Contains five governance processes (**figure 13**); within each process, evaluate, direct and monitor (EDM) practices are defined.
2. **Management**—Contains four domains, in line with the responsibility areas of plan, build, run and monitor (PBRM), and provides end-to-end coverage of IT. These domains are an evolution of the COBIT 4.1 domain and process structure. The names of the domains are chosen in line with these main area designations, but contain more verbs to describe them:
   - Align, Plan and Organise (APO)
   - Build, Acquire and Implement (BAI)
   - Deliver, Service and Support (DSS)
   - Monitor, Evaluate and Assess (MEA)

   Each domain contains a number of processes. The COBIT 5 process reference model is the successor of the COBIT 4.1 process model, with integrated Risk IT and Val IT process models. **Figure 13** shows the complete set of 37 governance and management processes within COBIT 5. The details of all processes, according to the process model described previously, are included in *COBIT 5: Enabling Processes*.

Figure 13—COBIT 5 Process Reference Model

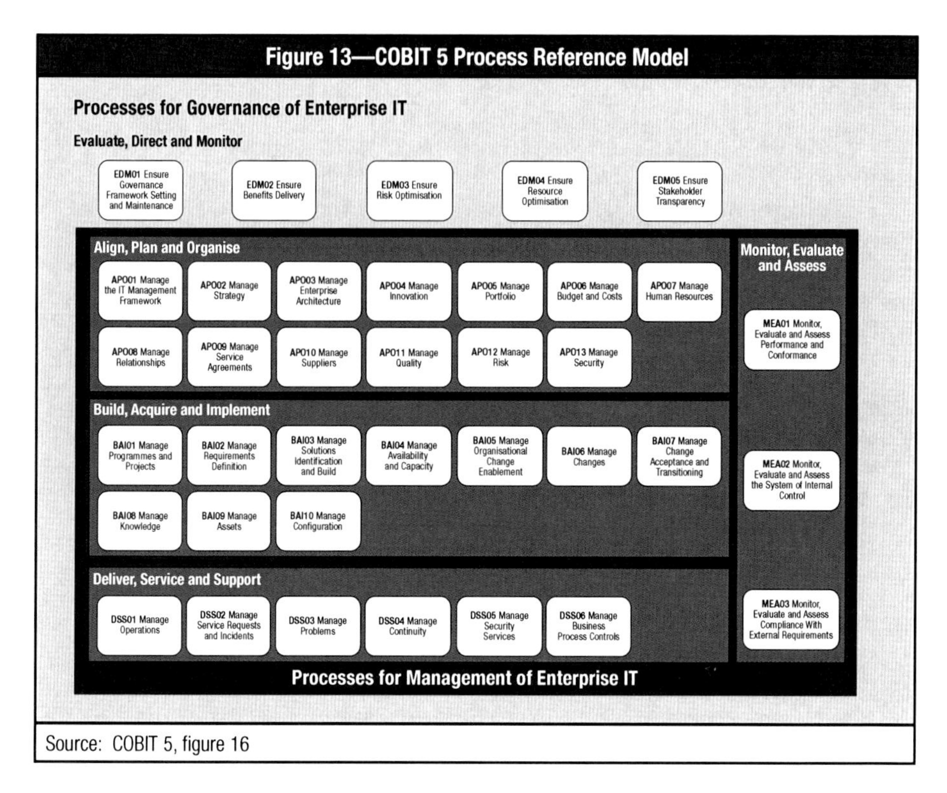

Source: COBIT 5, figure 16

COBIT 5-based Cloud Computing Assessment

A COBIT-based assessment provides the enterprise with a set of effective measures for gauging and controlling activities that are related to a cloud deployment or utilization. IT managers, working with application and data owners and administrators, can develop a comprehensive cloud computing-based assessment using:

- COBIT 5 process practices
- Detailed guidelines of the interrelationships of inputs and outputs to and from other COBIT 5 processes
- A responsible, accountable, consulted and/or informed (RACI) chart, which identifies likely risk assignments among organizational functions
- Specified process assessment capability model quantification

Appendix A provides the process practices from COBIT 5 and maps the process practices that are appropriate for enterprises to consider when choosing to utilize cloud computing services or when providing cloud computing services.

Governance Considerations

Establishing Stakeholder Needs for the Cloud Program

IT is most likely a strategic asset for an enterprise. When properly implemented, the cloud can be an extension of this asset. Proper implementation requires recognizing and establishing controls to offset known and future cloud risk. Before any of this can occur, the (cloud) governance committee needs to detail the enterprise's expectations and stakeholder needs for the cloud program.

Figure 14 details a suggested process sequence to obtain consensus across the enterprise. This governance process requires detailed interaction among all affected business groups. As noted in item 6 of this sequence, when the initially requested information is gathered from affected groups and assimilated, that information needs to be reviewed by other groups before adoption. It is possible that one business group's need for the cloud may impact a second group, but the cloud-seeking group may not understand the possible implications to the second group.

Figure 14—Process Sequence for Consensus to Use the Cloud

	Establishing Stakeholder Needs for a Cloud Program	*COBIT 5: Enabling Processes*
1	Identify the desired stakeholder needs beyond capabilities of current IT.	• APO02.02 Assess the current environment, capabilities and performance. • APO02.03 Define the target IT capabilities. • APO05.01 Establish the target investment mix. • APO05.03 Evaluate and select programmes to fund. • APO12.03 Maintain a risk profile. (Also COBIT 5 for Risk)
2	Define the opportunities envisioned for a cloud application.	• EDM02.02 Direct value optimisation. • APO02.04 Conduct a gap analysis. • APO02.05 Define the strategic plan and road map. • BAI01.02 Initiate a programme (to confirm expected benefits).
3	Quantify the gains envisioned in a cloud application.	• APO05 Manage portfolio. • APO06 Manage budget and costs. • BAI01.04 Develop and maintain the programme plan.
4	Identify the required cloud application processes to achieve stated goals.	• APO12.03 Maintain a risk profile. • BAI03.01 Design high-level solutions.
5	Identify the applicable compliance regulations.	• MEA03 Monitor, evaluate and assess compliance with external requirements.
6	Verify these values (for items 1 to 5) with all applicable stakeholders.	• BAI01.03 Manage stakeholder engagement.
7	Compare these goals for the cloud vs. traditional IT.	• BAI01.02 Initiate a programme (to confirm expected benefits).
8	Develop a detailed business case.	• EDM02 Ensure benefits delivery. • BAI01.04 Develop and maintain the programme plan.

Linking Stakeholder Needs to Cloud Computing With COBIT 5

The business aspect of COBIT 5 consists of linking stakeholder needs all the way to enabler goals, providing metrics and capability models to measure their achievement, and identifying the associated responsibilities of business and IT process owners.

COBIT 5's goals cascade allows a set of generic business and IT goals to be defined, which are based on stakeholder needs and provide a business-related and more refined basis for establishing business requirements and developing the metrics that allow measurement against these goals. Every enterprise uses IT to enable business initiatives, and these can be represented as enterprise goals for IT. COBIT 5 provides a set of matrices that maps generic stakeholder needs to the enterprise goals, enterprise goals to IT goals and IT goals to enabler goals.[14]

Following the methodology that is explained in the previous paragraph and with the help of the related matrices in COBIT 5,[15] an enterprise can implement GEIT for the cloud by starting with the stakeholder needs that are relevant for the enterprise, when making use of cloud capabilities and selecting the related enterprise goals (see **figure 15**).

Considering the general stakeholder needs for a cloud program (COBIT 5 chapter 1, compliance with relevant laws, regulations, contractual agreements and internal policies to protect information confidentiality, integrity and availability), it is fair to identify enterprise goals EG01, EG03, EG04, EG08 and EG10 in **figure 15** as particularly important in a cloud environment. The example drawing in **figure 15** demonstrates the correlation between cloud enterprise goals, IT goals and the COBIT 5 processes that must be implemented in a mature way to ensure the achievement of the goals. It is expanded in **figure 16**.

Figure 15—Goal Cascade for Cloud Computing

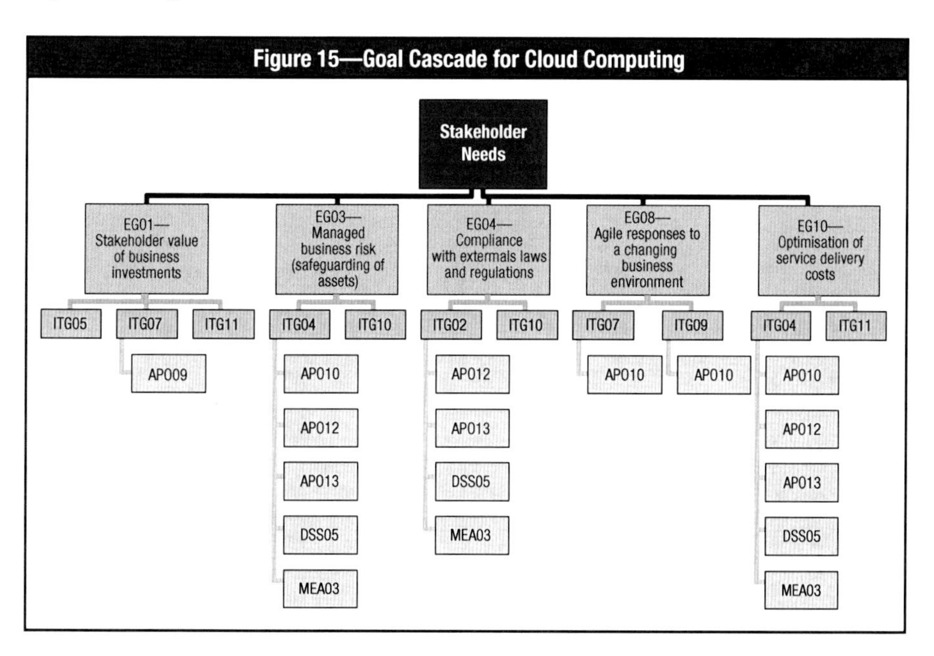

[14] ISACA, COBIT 5, USA, 2012, figures 22, 23 and 24
[15] *Ibid.*

Figure 16—Goal Cascade for Cloud Computing Expanded

EG	IT Goal		Process	
EG01	ITG05	Realised benefits from IT-enabled investments and services portfolio		
	ITG07	Delivery of IT services in line with business requirements	APO09 Manage Service Agreements	Align IT-enabled services and service levels with enterprise needs and expectations, including identification, specification, design, publishing, agreement, and monitoring of IT services, service levels and performance indicators.
	ITG11	Optimisation of IT assets, resources and capabilities		
EG03	ITG04	Managed IT-related business risk	APO10 Manage Suppliers	Manage IT-related services provided by all types of suppliers to meet enterprise requirements, including the selection of suppliers, management of relationships, management of contracts, and reviewing and monitoring of supplier performance for effectiveness and compliance.
			APO12 Manage Risk	Continually identify, assess and reduce IT-related risk within levels of tolerance set by enterprise executive management.
			APO13 Manage Security	Define, operate and monitor a system for information security management.
			DSS05 Manage Security Services	Protect enterprise information to maintain the level of information security risk acceptable to the enterprise in accordance with the security policy. Establish and maintain information security roles and access privileges and perform security monitoring.

Figure 16—Goal Cascade for Cloud Computing Expanded *(cont.)*				
EG	**IT Goal**		**Process**	
EG03 *(cont.)*	ITG04 *(cont.)*	Managed IT-related business risk *(cont.)*	MEA03 Monitor, Evaluate and Assess Compliance with External Requirements	Evaluate that IT processes and IT-supported business processes are compliant with laws, regulations and contractual requirements. Obtain assurance that the requirements have been identified and complied with and integrate IT compliance with overall enterprise compliance.
	ITG10	Security of information, processing infrastructure and applications		
EG04	ITG02	IT compliance and support for business compliance with external laws and regulations	APO12 Manage Risk	Continually identify, assess and reduce IT-related risk within levels of tolerance set by enterprise executive management.
			APO13 Manage Security	Define, operate and monitor a system for information security management.
			DSS05 Manage Security Services	Protect enterprise information to maintain the level of information security risk acceptable to the enterprise in accordance with the security policy. Establish and maintain information security roles and access privileges and perform security monitoring.
			MEA03 Monitor, Evaluate and Assess Compliance with External Requirements	Evaluate that IT processes and IT-supported business processes are compliant with laws, regulations and contractual requirements. Obtain assurance that the requirements have been identified and complied with and integrate IT compliance with overall enterprise compliance.
	ITG10	Security of information, processing infrastructure and applications		

Figure 16—Goal Cascade for Cloud Computing Expanded *(cont.)*

EG	IT Goal		Process	
EG08	ITG07	Delivery of IT services in line with business requirements	APO10 Manage Suppliers	Manage IT-related services provided by all types of suppliers to meet enterprise requirements, including the selection of suppliers, management of relationships, management of contracts, and reviewing and monitoring of supplier performance for effectiveness and compliance.
	ITG09	IT agility	APO10 Manage Suppliers	Manage IT-related services provided by all types of suppliers to meet enterprise requirements, including the selection of suppliers, management of relationships, management of contracts, and reviewing and monitoring of supplier performance for effectiveness and compliance.
EG10	ITG04	Managed IT-related business risk	APO10 Manage Suppliers	Manage IT-related services provided by all types of suppliers to meet enterprise requirements, including the selection of suppliers, management of relationships, management of contracts, and reviewing and monitoring of supplier performance for effectiveness and compliance.
			APO12 Manage Risk	Continually identify, assess and reduce IT-related risk within levels of tolerance set by enterprise executive management.
			APO13 Manage Security	Define, operate and monitor a system for information security management.
			DSS05 Manage Security Services	Protect enterprise information to maintain the level of information security risk acceptable to the enterprise in accordance with the security policy. Establish and maintain information security roles and access privileges and perform security monitoring.

Figure 16—Goal Cascade for Cloud Computing Expanded *(cont.)*				
EG	**IT Goal**		**Process**	
EG10 *(cont.)*	ITG04 *(cont.)*	Managed IT-related business risk *(cont.)*	MEA03 Monitor, Evaluate and Assess Compliance with External Requirements	Evaluate that IT processes and IT-supported business processes are compliant with laws, regulations and contractual requirements. Obtain assurance that the requirements have been identified and complied with and integrate IT compliance with overall enterprise compliance.
	ITG11	Optimisation of IT assets, resources and capabilities		

Appendix B contains an example assurance program for cloud computing. It provides a template for enterprises that are seeking assurance over all of their cloud computing activities. One of the first steps of this assurance template is establishing this goal cascade.

Governance Outcomes

By initiating a governance program, enterprise leaders can identify, at a high level, cloud computing's benefits and risk. The COBIT 5 family frameworks provide guidance on drilling down into the detailed aspects of each broad risk and value area, while allowing final analysis results to percolate back up to high-level issues.

When employing the described governance, methodologies and tools, decisions that were previously based on subjective rationale are executed in a better manner. Applying the COBIT 5 family of publications provides an excellent means to determine the kind of service and deployment model that the enterprise needs, and whether CSP offerings align with enterprise business expectations (valuations) and business risk tolerance levels. The following section takes a closer look into the next steps.

The Path to the Decision and Beyond[16, 17]

This section provides practical guidance on how to consider a potential decision to go to the cloud. Two decision trees are outlined to help prospective cloud users decide whether they should move assets to the cloud and, if so, which service and deployment models are best for their enterprise. In this context, the following approach can be taken:

Step 1. Preparation of the internal environment
Step 2. Selection of the cloud service model
Step 3. Selection of the cloud deployment model
Step 4. Selection of the cloud provider

[16] ISACA, *Security Considerations for Cloud Computing*, USA, 2012, pp. 35-52, *www.isaca.org/cloud-security*
[17] "The Path to the Decision and Beyond" is copied from chapter 4 in *Security Considerations for Cloud Computing.*

However, the challenge does not end after step 4. Even if the enterprise has decided to go to the cloud based on these steps and the enterprise trusts the CSP, there are still a number of questions that must be answered. These questions will be addressed through the mitigating actions discussed in chapter 3. These mitigating actions can be translated into a checklist that management should use in deciding to move to the cloud.[18]

In addition to this publication, practical guidance on implementing good practices relative to IT governance can be found in ISACA's publication *COBIT 5 Implementation*, which includes an implementation tool kit containing a variety of resources that are continually enhanced to reflect current trends. Its content includes:
- Self-assessment, measurement and diagnostic tools
- Presentations aimed at various audiences
- Related articles and further explanations

Step 1. Preparation of the Internal Environment

Besides selecting deployment and service models, an enterprise must do some preparatory work to make a migration to the cloud possible.[19] All IT dimensions should be taken into account when defining the project scope and project plan. The COBIT 5 enablers discussed in the previous sections of this chapter provide practical guidance when looking into the different aspects:
- **Principles, Policies, and Frameworks**—Which security policies apply within the enterprise? Which regulatory restrictions apply to the enterprise and to any locations where a CSP might reside?
- **Processes**—How will moving to the cloud influence the enterprise's processes? Which processes depend on assets that could move to the cloud? Are these processes considered to be critical for the business?
- **Organisational Structures**—How will the relationship with the CSP be managed? How are roles and responsibilities defined?
- **Culture, Ethics and Behaviour**—How will change within the enterprise be managed? How can an information culture be imposed upon the CSP?
- **Information**—Which assets are considered for cloud computing? The enterprise should classify its assets into categories for an optimal selection of cloud arrangements. Generally, data can be classified as public, restricted, for internal use, secret and top-secret. A data life cycle process can also be defined.
- **Services, Infrastructure and Applications**—Which service capabilities are expected of the CSP? How will performance be measured? How will issues be reported?
- **People, Skills and Competencies**—Which skills and competencies are required to manage the assets of the enterprise? Does the enterprise wish to keep these in-house after a decision has been made to move to the cloud?

In addition to these considerations, the enterprise's decision to migrate to the cloud must take into account a consistent business case and an evaluation of the costs and benefits related to the move to the CSP.

[18] ISACA, *Security Considerations for Cloud Computing*, USA, 2012, *www.isaca.org/cloud-security*
[19] Commercial analysis must, of course, be done, but it is out of scope for this publication.

Calculating Cloud Computing ROI[20]

Return on Investment (ROI) is one of several financial metrics available to estimate the financial outcome of business investments. This calculation considers both the costs of an investment and its expected gains, and yields an estimate of how favorable the investment will be. To calculate ROI (simple ROI), the result of subtracting the cost of an investment from the gain (return) of the investment is divided by the cost of the investment and the result is expressed as a percentage or ratio (**figure 17**). In most cases, a ratio greater than 0 means the return is greater than the cost, therefore the investment may be considered beneficial[21] (how beneficial depends on the enterprise's investment objectives or its corporate standards).

Figure 17—Formula to Calculate Simple ROI

$$\text{ROI} = \frac{(\text{Gain From Investment} - \text{Cost of Investment})}{\text{Cost of Investment}}$$

For example the ROI for a new cloud based application (SaaS) that is expected to have an investment of $600,000 over a period of five years and provide benefits (cost savings and new revenue) of $900,000 over the same period of time will yield a return of 50 percent.

$$\text{ROI} = \frac{\$900{,}000 - \$600{,}000}{\$600{,}000} = 50\%$$

ROI calculations used as the only financial measurement for decision making do not help predict the likelihood of realizing the return or the risk involved with a particular investment. Ideally, the enterprise will use multiple financial metrics (e.g., TCO, net present value [NPV], internal rate of return [IRR], payback period) in considering whether to adopt cloud computing services. TCO is different from ROI because it accounts only for the cost associated with an acquisition for its entire life span or a period of time determined for the calculation. NPV compares anticipated benefits and costs over a predetermined time period using a rate that helps calculate the present value of future cash flow transactions. IRR is a variant of NPV used to find the discount rate that would make the NPV of the investment equal to zero. TCO, NPV and IRR are more significant and complex calculations; therefore, they require additional data and variables for their calculation. ROI's simplicity makes it a more popular term to use in marketing materials and project analysis.

For investments that have clear and quantifiable benefits and costs that are easily known, the ROI calculation is simple. However, for more complex investments such as cloud computing services, the ROI calculation can be complex and misleading. Generating a meaningful result is dependent on accounting for all quantifiable

[20] ISACA, *Calculating Cloud ROI: From the Customer Perspective*, USA, 2012, *www.isaca.org/Knowledge-Center/Research/ResearchDeliverables/Pages/Calculating-Cloud-ROI-From-the-Customer-Perspective.aspx*

[21] Schmidt, Marty J.; "Encycopedia of Business Terms and Methods: Return on Investment," Solution Matrix Ltd., USA, 2011

variables and defining a clear and consistent time period. Intangible benefits and risk may not be included in the calculation unless the business is able to assign a value based on historical or statistical data. Investments based solely on business objectives may be better justified using a business case supported by multiple financial metrics.

The cloud promises a range of benefits that include the ability to shift cost from capital to operational, lower overall cost, greater agility and standardization, the ability to shift IT resources to higher-value-added activities, improved employee satisfaction, and competitive advantage. Some of these benefits are quite subjective and, therefore, are difficult to include in financial (mathematical) calculations.

Cloud Benefits

Figure 18 describes the most common benefits promised by cloud computing supporters. The benefits are grouped into two categories: tangible (quantifiable) and intangible (strategic) benefits.

Figure 18—Cloud Benefits

Benefit	Description
Tangible	
Cost reduction	Computing cost is shifted from capital expenditure to operational cost because the cloud provider supplies the underlying infrastructure as part of the service bundle. In addition, the cloud promises cost reduction in the following areas: • Labor—IT system administration hours/headcount • Application software (SaaS only) • Licensing purchase and maintenance • Technical support and user support • Maintenance (upgrades, updates, patches, etc.) • Hosting (physical building, power, cooling, etc.)
Enhanced productivity	User mobility and ubiquitous access can increase productivity. Collaborative applications increase productivity and reduce rework.
Optimized resource utilization	Enterprises use only the computing resources they need, thus reducing system idle time waste.
Improved security/compliance	Cloud providers may offer robust security controls as a market differentiation.
Access to skills and capabilities	Customers benefit from top-notch skills and capabilities while avoiding employment cost (recruiting, salary, benefits, training, etc.).
Scalability	On-demand provisioning or computing resources eliminate the cost of capacity planning.
Agility	Agility contributes to cost reduction and productivity enhancement due to faster provisioning of systems: • Faster application deployment (SaaS) • Faster application development/testing (PaaS)

Figure 18—Cloud Benefits *(cont.)*	
Benefit	**Description**
Tangible *(cont.)*	
Customer satisfaction	Effective utilization of cloud applications can increase collaboration between the enterprise and its customers or reduce response time to customer inquiries.
Reliability	Cloud providers have redundant sites that can address business continuity and disaster recovery in a more efficient manner.
Performance	Better performance and up-time can result from continuous and consistent operations monitoring by the cloud provider.
Intangible	
Avoidance of missed business opportunities	A cloud application (SaaS) may be the critical element to land a new business or expand into new markets.
Focus on core business	IT resources can be allocated to support core business functions.
Employee satisfaction/innovation	Mobility and faster performance can improve employee satisfaction and boost innovation.
Collaboration	Real-time collaboration can increase quality and innovation.
Risk transfer	Some risk can be transferred to the CSP (e.g., security breaches, data loss, disaster recovery); this could represent a tangible or intangible benefit.

Cloud Costs

Cloud solutions include many elements beyond the obvious hardware and software costs. There are three types of costs: start-up (upfront costs), operational (recurring costs) and one-time (change or termination costs). **Figure 19** describes the most common of these cost types.

Figure 19—Cloud Costs	
Cost	**Description**
Upfront costs	
Technical readiness	Some investment in bandwidth may be necessary to accommodate the new demand for network/Internet access. Other infrastructure components might need to be upgraded to integrate with cloud services.
Implementation	Professional services may be needed for managing the transition to the cloud.
Integration	Professional services may be needed for integrating in-house and cloud services.
Configuration/customization	This applies to customer-based configuration for SaaS applications.
Training	IT resources may require training to manage cloud vendors and services. Users may need training on new applications.

Figure 19—Cloud Costs *(cont.)*

Cost	Description
Upfront costs *(cont.)*	
Organizational change	Processes may require some reengineering to accommodate cloud-specific needs (e.g., change management, resource utilization monitoring, user access provisioning, internal audit).
Recurring costs	
Subscription fees	These will comprise agreed-on periodic fees (monthly, quarterly, yearly) for the use of cloud services.
Change management	These may comprise cost associated with the change management process and any cost incur when requesting system changes.
Vendor management	These are costs associated with monitoring CSP activities, contract management, SLAs monitoring and enforcement, or any other activity geared to manage service delivery and evaluation.
Cloud coordination	For enterprises running more than one cloud service, a cloud coordination group is necessary to ensure integration and consistency.
End-user support and administration	Some of these costs will be part of the subscription fee, but some may be retained by the enterprise.
Risk mitigation	Countermeasures will need to be implemented to control any risk introduced by cloud computing.
Downsize/upsize	Unless otherwise specified in the contract, some vendors may charge for downsizing or upsizing computing resources.
Termination costs	
Revert to on premises or transfer to a different provider	The enterprise may need to revert to an in-house model when/if new regulations or economic problems render the cloud impractical. Some of the possible costs are: • Extracting data out of the cloud and validating data accuracy and completeness • Cost to sanitize or shred data from cloud storage and processing hardware • Configuration and provisioning in-house systems to replace cloud services • Penalties for early termination • Reallocation or recruitment of IT resources to support services being reverted • Reallocation or procurement of physical resources to host services being reverted

Business Challenges to Consider

There are business challenges when using the cloud. **Figure 20** describes common challenges that should be considered when evaluating cloud services.

Figure 20—Cloud Challenges

Challenge	Description
Incompatibility	Cloud services may not be compatible with the existing IT infrastructure or specific systems that must be integrated.
Uptime	Cloud vendors may not be able to guarantee agreed-on uptime. In addition, uptime may be impacted by other factors including the customer's Internet service providers.
Performance	Multitenant models can degrade performance over time if capacity is not properly planned. Internet speed can also negatively impact performance.
Security	Cloud computing represents traditional and new risk that must be accounted for and mitigated accordingly (either by the CSP or the customer)
Compliance	The ubiquitous and abstract nature of the cloud could cause an enterprise's transition from compliance to noncompliance without any notice.
Pay-as-you-go	The enterprise must implement controls to avoid overage charges incurred when systems stay connected after a demand spike is over.
Lock-in (hardware or vendor)	Customers may become locked into a specific technology or a specific cloud vendor, which can prevent portability.
Cloud consumerization	Business units may be able to procure cloud services without involving IT. To prevent this situation, the enterprise must adapt its governance framework to control cloud services procurement.
Limited customization (Black Box)	Cloud applications may not be customized every time the business process changes, making the business process a "Black Box" due to cost associated with each modification or application limitations.

The previous tables are not all-inclusive. They list the best-known benefits, costs and challenges, and are meant to kick-start cloud ROI analysis. Users must include values relevant to the project under analysis and remove nonapplicable values.

Decision making around use of cloud services can be complex and estimating the ROI is a critical part of ensuring that the path taken is the right one. Following are some key points to bear in mind:

- Estimating ROI does not need to be complex. After all, it is just an estimate. A simple, but effective, ROI calculation enables the enterprise to support an investment decision and measure whether or not the expected cost and benefits occur. An overly complex calculation can make it hard to understand why a decision was made and/or measure its effects, essentially defeating the purpose of doing one in the first place.

- Cloud is not right for every organizational need. The type of cloud service selected is critical, as is how it is managed. Thinking strategically about benefits, costs and risk is paramount and must be done up front, before any contract is signed.
- There can be many hidden costs that are not obvious from the provider's fee schedule. For example, while there may not be any up-front service provisioning costs from the provider, the time and effort in migrating existing systems into the cloud can be expensive. The same can be said of pulling systems or data back in-house or porting them to another provider. The lesson is that selecting the right cloud service can result in cost savings, but selecting the wrong one can be very expensive indeed.
- It is far easier and cheaper to change a decision (e.g., different service model or provider) when it is still on the drawing board or perhaps in proof-of-concept stage. It can be far more difficult and expensive when the service is up and running, interfacing with other systems and processes and using live customer data. With so many cloud service options available, the time the enterprise spends considering the respective ROIs and selecting the best fit for its needs is time well spent.

A Practical Approach to Measuring Cloud ROI

There are many possible ways to estimate cloud ROI and no single approach will suit all situations. Selecting the best one for a particular case depends on several factors, including what the key business drivers are for moving to the cloud (increasing revenue vs. reducing costs), the enterprise's approach to preparing and assessing business cases (focus on tangibles vs. intangibles), and where the enterprise is in the business growth/maturity cycle (start-up vs. mature business).

The three-phase approach outlined in this book is designed to suit an enterprise that has reasonably mature operations (i.e., existing systems and business processes) and is considering moving to the cloud primarily to achieve cost savings. The concepts described can also be applied to other scenarios; some steps may need more or less emphasis to suit the circumstances.

A few suggestions for maximizing this approach follow:

- **Focus quickly on the optimal cloud solution**—There are so many alternatives that sifting through them all could take forever. Starting with an initial/baseline model and then iteratively identify the one best suited to the enterprise's needs (cost, risk, compliance, etc.) can make the selection process faster and more effective.
- **Make an "apples to apples" comparison**—Evaluate a holistic and comparable set of costs for both the as-is and to-be alternatives addressing the common problem of not being able to make a fair comparison between two solutions that are potentially very different (either comparing two different cloud solutions or comparing cloud to traditional IT). Measuring monetary values in a consistent manner increases ROI accuracy and reliability.
- **Stay within the enterprise's risk tolerance**—Perform a risk assessment of both the as-is and to-be options to help ensure that the solutions being compared are within the enterprise's risk tolerance and the costs of mitigating unacceptable risk are factored into the calculations. Knowing the enterprise's risk appetite before the calculations begin is a must.

Appendix G outlines the three phases and suggested questions to address each step.

ROI Framework in Practice

Figure 21 illustrates the progression to a decision.

Figure 21—Cloud Decision Path

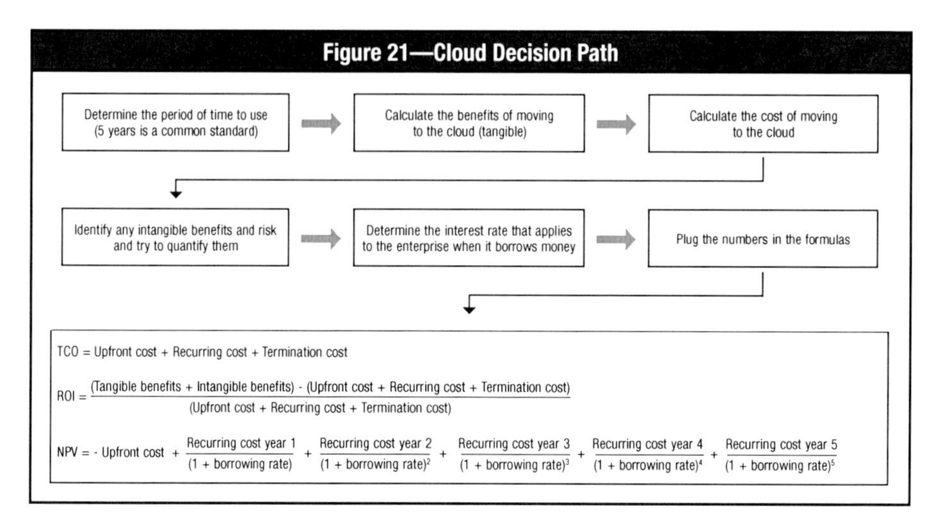

Appendix G describes these formulas in more detail.

Following are some points to consider when applying this approach:

- **Phase 1:**
  - A good reason to start by evaluating a baseline cloud model that is simple and cost-effective is to facilitate conducting a proof of concept to better understand the model's features, benefits and risk. With low sign-up and operating costs, many public cloud solutions can be very helpful for this purpose.
  - If the enterprise is sure that it has already identified its optimal cloud option, it may be possible to skip the steps in phase 1 related to evaluating an initial/baseline model. However, if the ROI of the optimal model has not yet been estimated, the question may be raised as to how it was determined that it is the optimal model ("the chicken or the egg" problem).
  - Gaining a firm understanding of the risk related to cloud services can be challenging due to the wide variety of services offered, the lack of transparency around controls and the difficulties in comparing across providers. While original investigation and research is effective, the following tools and references may be helpful in the process:
    - Links to ISACA's cloud resources can be found on page *www.isaca.org/Knowledge-Center/Research/Pages/Cloud.aspx.*
    - ISACA's *Security Considerations for Cloud Computing*—Practical guide to assess cloud risk and assists in determining which cloud model is most propitious to satisfy enterprise needs
    - Cloud Security Alliance's (CSA) *Security Guidance for Critical Areas of Focus for Cloud Computing*—Detailed guidance on principles and practices for cloud computing security

- NIST SP 800-146 *Cloud Computing Synopsis and Recommendations*
- CSA's Security, Trust & Assurance Registry (STAR)—A free public registry of security controls provided by various cloud computing offerings
- European Network and Information Security Agency's (ENISA) *Cloud Computing: Benefits, Risks and Recommendations for Information Security*—A well thought-out and highly detailed paper on cloud risk and benefits

- **Phase 2**—This phase may be relatively straightforward, depending on how much documentation and analysis exists of the current service model and associated costs. But much depends on whether or not a full assessment of the current model has been completed. If not, and there are unknown areas of risk, the enterprise could be significantly underestimating as-is costs and/or to-be benefits.
- **Phase 3**—Many enterprises moving to the cloud are redirecting a significant portion of IT operating cost savings toward managing cloud-related risk and management. This is because the cloud is introducing new types of risk, and the methods to manage that risk can be quite different from the approaches used for traditional IT (e.g., vendor management, change management, usage management). This may be reflected in as-is and to-be costs.

After the preparation of the internal environment, the following step is to look into the selection of a cloud service and deployment model. The flowcharts presented in steps 2 and 3 are examples that illustrate some of the factors the enterprise should consider in order to determine which cloud service model and which cloud deployment model could best suit the enterprise needs.

While the questions were chosen very carefully to accommodate a maximum of enterprise needs, the flowcharts only serve as an example of what type of questions should be taken into consideration. Questions can be added or adapted to better serve individual enterprise needs.

Step 2. Selection of the Cloud Service Model

The most common technical reason not to move to the cloud is that the cost of customization outweighs the benefits of the cloud solution.

The decision tree presented in **figure 22** is an example of a method that can be used to help the enterprise determine which service model best serves its business needs. The decision tree may lead to a decision to migrate to the cloud, but it may also suggest that the cloud is not the optimal solution for the enterprise and that other solutions, such as outsourcing or keeping the process in-house, may be more viable options.

The cloud deployment model addresses potential risk and its mitigation, while the service model is more focused on a technical solution. This explains why not all possible outcomes in the decision tree end in a cloud service model.

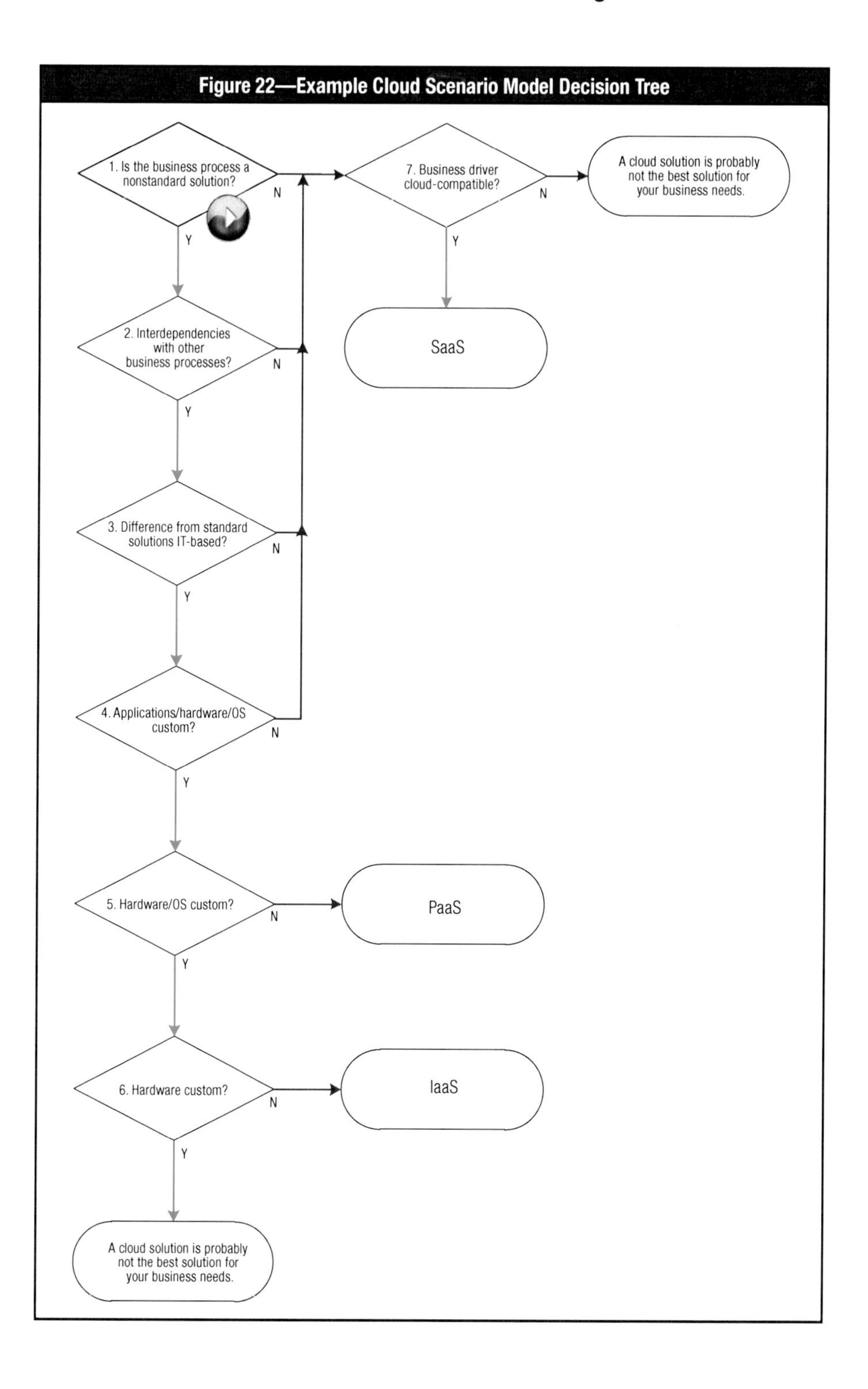
Figure 22—Example Cloud Scenario Model Decision Tree
1. Is the business process a nonstandard solution?
N
Y
7. Business driver cloud-compatible?
N
Y
A cloud solution is probably not the best solution for your business needs.
SaaS
2. Interdependencies with other business processes?
N
Y
3. Difference from standard solutions IT-based?
N
Y
4. Applications/hardware/OS custom?
N
Y
5. Hardware/OS custom?
N
Y
PaaS
6. Hardware custom?
N
Y
IaaS
A cloud solution is probably not the best solution for your business needs.

Figure 23 provides a breakdown of the cloud service model decision tree.

Figure 23—Breakdown of the Example Cloud Service Model Decision Tree

Answer	Explanation
1. Is the business process a nonstandard solution?	
Yes	If the business process uses nonstandard solutions, then a further drilling down is needed to determine whether the business process is suitable for a cloud solution.
No	If a standard solution is used, then the transition to the cloud is relatively easy and the benefits of adopting a cloud solution will most likely be high.
2. Interdependencies with business processes?	
Yes	If there are interdependencies with different business processes, then any alteration to one of these processes could mean a change to the application implemented in the cloud.
No	If there are no interdependencies, then changes will not be required. The chosen cloud solution will, therefore, be independent.
3. Difference from standard solutions IT-based?	
Yes	While interdependency may implicate a change in the IT infrastructure, it is not always a necessity. If interdependency does implicate such a change, however, the cloud application will need to be changed. This fact will largely influence the decision for a cloud service model. Thus, it is important to outline the differences between the current solution and the standard solution provided by a CSP.
No	If there are no differences between the IT solutions, then the standard offerings of a CSP will adequately address the business needs.
4. Application/hardware/OS custom?	
Yes	Once it is established that there is indeed a gap between the business needs and the cloud service offerings, it is important to define the level on which the difference is situated.
No	If the differentiation is situated in the configuration of standard applications, then cloud offerings will fulfill the business needs.
5. Hardware/OS custom?	
Yes	After establishing that the difference is not within the application, it is important to establish whether the differentiation is found on the OS level or the physical hardware platform. The answer will alter the possibility for cloud adaptation.
No	If the differentiation can be done on application level, no further drill-down is needed.
6. Hardware custom?	
Yes	After establishing that the differentiation is located on the physical level, a cloud solution is very unlikely. CSPs are oriented toward standardization within their domain; providing custom hardware is not one of their typical offerings. While a CSP can undoubtedly provide custom hardware platforms, the high cost and the CSPs relative lack of experience in the custom platform eliminate the cloud as a viable solution.
No	If the differentiation can be done on the OS level, no further drill-down is needed.

Figure 23—Breakdown of the Example Cloud Service Model Decision Tree *(cont.)*

Answer	Explanation
7. Business driver cloud-compatible?	
Yes	Viable business drivers for the cloud decision include: • Reduce medium- and/or long-term total cost of ownership (TCO). • Improve cash flow by decreasing investments. • Shift from capital expenditures (CAPEX) to operating expenditures (OPEX). • Improve Quality of Service (QoS) and/or SLAs. • Gain access to functionality and/or domain expertise.
No	While there may be no technical constraints to adopting the cloud as a solution, it is possible that the business drivers are, in fact, not cloud-compatible. Adopting a cloud solution requires a mid- to long-term vision. Therefore, the cloud cannot be used as a solution to cut costs immediately.

Step 3. Selection of the Cloud Deployment Model

While there are four common cloud deployment models, the decision tree presented in this section focuses on deciding between a private or public cloud. Hybrid cloud or community cloud are deployment models that arise for consideration when evaluating several cloud solutions that are present in one enterprise or collection of enterprises.

A hybrid cloud is most commonly used when there is a data classification system in place and the decision is made to use different deployment models for different data classifications (e.g., a private cloud model for human resources (HR) data and a public cloud for storage of publications).

The same goes for a community cloud. A community cloud is created when several allied companies or enterprises decide to move to the cloud together. Either the community as a whole decides to create a common infrastructure platform for all to use (common reasons being the ease of sharing information and cost reduction), or one member or sponsor provides the necessary infrastructure that is used by the community.

The example decision tree (shown in **figure 24**) also offers the options of not going to the cloud at all or considering alternatives to the cloud. This decision (among others) is made when the data or the process is too critical or contains so much sensitive or business-critical data that the risk of going to the cloud outweighs the benefits.

Note: When the situation addressed in the question does not occur or when it can be adequately covered by technical means, policies or contracts, the question should be answered affirmatively.

Figure 24—Example Cloud Deployment Model Decision Tree

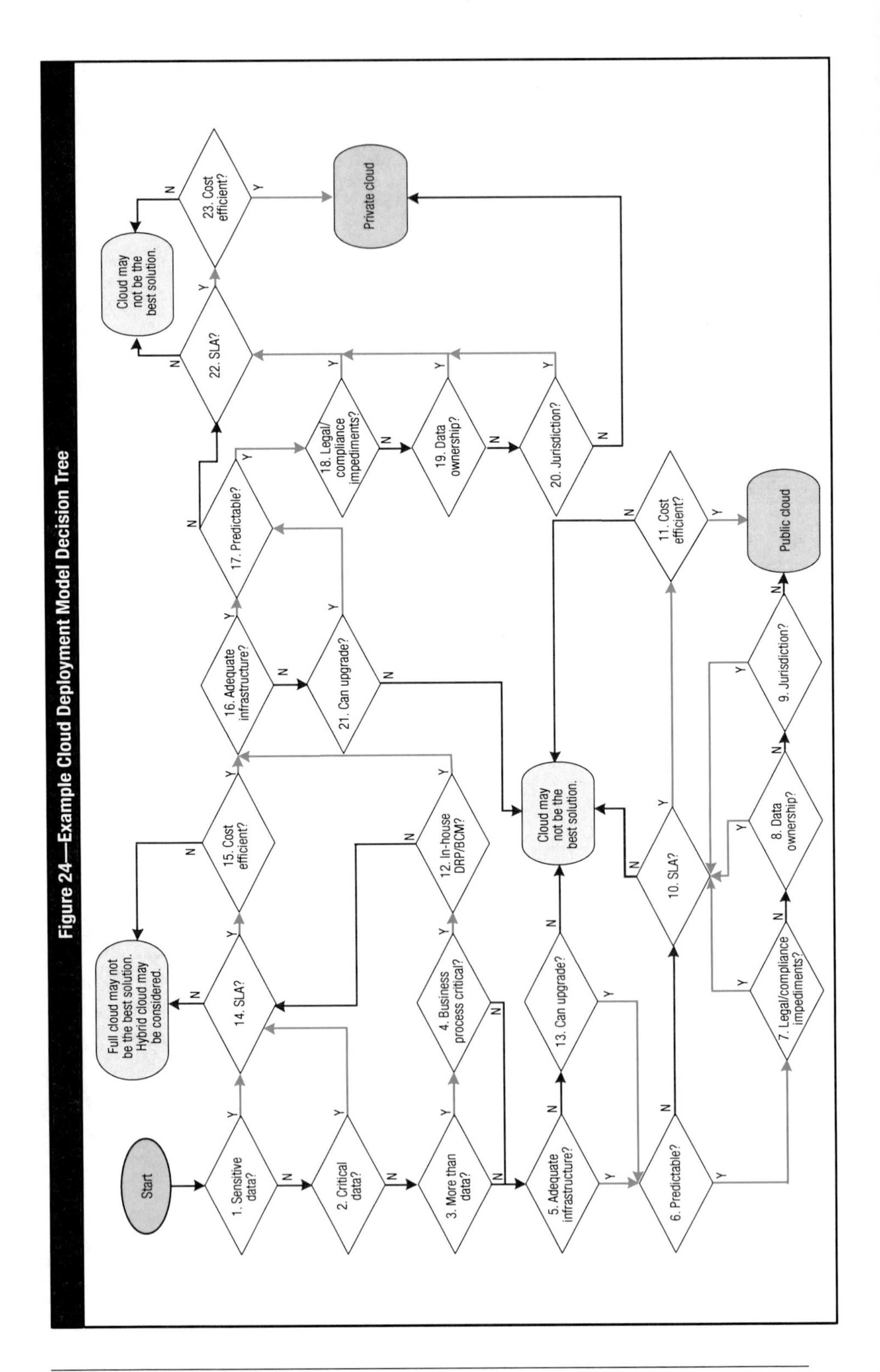

Figure 25 provides a breakdown of the cloud deployment decision tree.

Figure 25—Breakdown of the Example Cloud Deployment Model Decision Tree	
Answer	**Explanation**
1. Sensitive data?	
Yes	When considering a move to a cloud infrastructure it is very important to be aware what data are to be released to the cloud. It is impossible to envision all potential risk and threats; however, data of a sensitive nature can be placed in the cloud when the necessary controls to protect them are in place and work effectively.
No	If data are not sensitive or if not data upload to the cloud is required, the first steps toward the cloud are taken
2. Critical data?	
Yes	Critical data can be: • Blueprints • Formulas • Trade secrets • Any information absolutely necessary for the enterprise to operate Critical data can be placed in the cloud when necessary controls to protect them are in place and working effectively. It is important to note, however, that some of these controls can be expensive and complex, which may increase the cost of moving to the cloud.
No	Noncritical data can be easily placed in the cloud.
3. More than data?	
Yes	In almost all cases the decision to move to the cloud is not restricted to solely the data. Data in the cloud are often needed to run application or as part of business processes.
No	If the decision to move to the cloud is limited to data only, the next step is to evaluate the enterprise's readiness for this move.
4. Business process critical?	
Yes	To make a sound decision, it is imperative to determine whether data and applications hosted in the cloud support critical business processes. This information will help determine the requirements that the cloud solution must satisfy.
No	When a business process or supporting application is not considered critical, it may be easier to move to the cloud.

Figure 25—Breakdown of the Example Cloud Deployment Model Decision Tree *(cont.)*	
Answer	**Explanation**
5, 16. Adequate infrastructure?	
Yes	A move to the cloud is a step toward reducing the enterprise's IT infrastructure; however, proper planning is needed prior to adopting a cloud solution. Some things to consider as part of the readiness assessment include: • Connectivity to the CSP (bandwidth, redundancy) • Network security (data encryption during transfer) • Integration between cloud and non-cloud systems • User connectivity (bandwidth to the desktop or mobile devices)
No	If it is determined that the current enterprise infrastructure is not ready to integrate with the cloud, the next step is to determine whether the business needs are greater than the cost to upgrade (feasibility analysis).
6. Predictable?	
Yes	As part of the readiness assessment the enterprise must determine how business processes function and mature. This information can help anticipate capacity fluctuation (up or down) that must be part of the contract with the CSP.
No	When the enterprise cannot anticipate capacity fluctuations, further analysis may be needed. Flexibility and scalability are two of the cloud characteristics that make it attractive—a flexible SLA may be the solution until the enterprise has more refined requirements.
7, 18. Legal/compliance impediments?	
Yes	There may be legal or compliance reasons why data or certain business functions cannot be moved to the cloud. It is important for the CSP to implement the necessary controls to ensure the enterprise's legal and compliance continuity. The CSP must be able to provide proof of compliance as reported by a neutral audit or control body. Identification of legal or compliance limitations must be addressed during contract negotiations to stipulate the enterprise's expectation and how they will be satisfied.
No	If the enterprise does not have any legal or compliance impediments, the next steps to move to the cloud can be taken.
8, 19. Data ownership?	
Yes	The contract with the CSP should clearly stipulate that the enterprise is, and will remain, the data owner. It is equally important that this ownership be maintained throughout the entire data life cycle. Therefore, the contract should also outline the requirements to dispose of data in an adequate manner when the enterprise deems necessary. If data ownership cannot be properly established, the enterprise may choose to move only nonsensitive and noncritical data.
No	If the enterprise can clearly define data ownership during contract negotiations, the next steps to move to the cloud can be taken.

Figure 25—Breakdown of the Example Cloud Deployment Model Decision Tree *(cont.)*	
Answer	**Explanation**
9, 20. Jurisdiction?	
Yes	Even though data ownership resides with the enterprise, local and international laws often forbid the transfer or certain data to countries that have conflicting laws or regulations. Therefore, it is important for the enterprise to know the location of the CSPs data storage facilities and data processing centers to prevent legal infractions. If is advisable for the enterprise to include in the contract with the CSP the necessary clauses requiring the CSP to limit service locations to those approved by the enterprise.
No	If the enterprise does not have jurisdiction limitation, the cloud may be a proper solution.
10, 14, 22. SLA?	
Yes	The enterprise must determine in advance the terms that will be included in the SLA keeping in mind that strict or complex SLAs could result in higher maintenance cost. Some of the terms that should be negotiated and documented in the SLA include: • Availability • Response time for additional computing resources requests • Response time for incidents • Backup policies • Data retention and disposal policies and procedures • Path management • Security controls • Recovery and continuity objectives • Controls to satisfy legal and compliance requirements
No	If an adequate SLA cannot be agreed on, moving to the cloud could pose an unacceptable level of risk. If the cost of the SLA is greater that the business driver, the cloud solution may not be the best solution.
11, 15, 23. Cost efficient?	
Yes	Two of the principal goals of moving to the cloud are becoming more cost effective and being able to react more quickly and inexpensively to changing situations.
No	Unless the business driver is greater than the cost, an expensive solution may not be the right option.

Figure 25—Breakdown of the Example Cloud Deployment Model Decision Tree *(cont.)*	
Answer	**Explanation**
12. In-house DRP/BCM?	
Yes	This question may already be addressed in the SLA, but the enterprise must still be ready to consider additional DR and BC plans. A disaster occurring within the CSP is likely to cause an impact on the enterprise's operations. For example, routes will change and entry points will be altered, causing delays in operations. If a disaster takes place within the enterprise, maintaining or reestablishing connectivity with the CSP should be a critical part of the recovery efforts. Enterprises whose data reside only on the cloud should create backups to their own premises to retain recovery and continuity capabilities even if the CSP is completely offline.
No	Relying solely on the DRP/BCM capabilities of the CSP can expose the enterprise to extended business outages; however, if the cost of having an in-house DRP is greater that the business driver, the enterprise may address this question in a more strict SLA.
13, 21. Can upgrade?	
Yes	If it is determined that the current enterprise infrastructure is not ready to integrate with the cloud, the next step is to determine whether the business needs are greater than the cost to upgrade (feasibility analysis).
No	If the cost to upgrade the current infrastructure is greater that the business needs, the cloud may not be a solution yet.
17. Predictable?	
Yes	As part of the readiness assessment the enterprise must determine how business processes function and mature. This information can help anticipate capacity fluctuation (up or down) that must be part of the contract with the CSP.
No	When the enterprise cannot anticipate capacity fluctuations, further analysis may be needed. Flexibility and scalability are two of the cloud characteristics that make it attractive—a flexible SLA may be the solution until the enterprise has more refined requirements.

Note: The decision trees included in this book are for illustration purposes only.[22] The enterprise should evaluate the internal and external environment to determine what questions should be included when trying to select a service and deployment models. In addition cloud benefits and challenges must be clearly included in any business case and financial calculations as described in the "calculating ROI" section.

Step 4. Selection of the Cloud Service Provider

Once the best-suited cloud flavor is established, it is key to find the best-suited CSP. Matching the correct CSP to the enterprise needs could prove challenging. The key is to find the CSP that bests serves the business needs while minimizing potential risk.

While there are many suitable smaller CSPs, it is important to choose an established CSP with the proper references. Established CSPs will potentially be more experienced with running cloud infrastructures, adapting to change and generally faster in responding to an incident or threat, thus being able to maintain stability more efficiently. One must keep in mind that a larger CSP will also be stricter toward its offerings, which will greatly reduce the possibility of fine-tuning services not fully compatible with the enterprise needs.

[22] ISACA, *Security Considerations for Cloud Computing*, USA, 2012, *www.isaca.org/cloud-security*

The following characteristics can influence the selection of a particular CSP depending on the objectives and goals established as part of the governance framework:

- **Vendor expertise**—The enterprise needs expert knowledge or a broader perspective and experience with similar enterprises to effectively and efficiently handle certain activities.
- **Vendor capacity**—The enterprise does not have the resources to handle the work related to a specific product or service. The vendor can supply the resources to support the entire operation or to supplement in-house resources.
- **Vendor assuming risk**—The enterprise outsources activities to leverage a vendor's experience with operational risk and corresponding risk mitigation services. However, the accountability for the adequate performance of those activities can never be delegated and stays with the enterprise.
- **Vendor leveraging scale**—Vendors can offer services at a lower cost because working for multiple customers allows vendors to leverage scale.

Vendor location (storage and processing centers) is of key importance for legal reasons. Laws are applicable locally, which can result in a multitude of different law systems applicable to data or business processes stored in the cloud.

The most important applicable laws are:

- **Local law of the enterprise**—Storing data in the cloud does not waive the enterprise from the legal obligations it has toward its customers, employees and shareholders. Therefore, local laws will still apply even if data are stored abroad.
- **Local law of the CSP**—Since the CSP has legal obligations toward its customers, meaning the enterprise, the local laws of the CSP may become applicable to its data centers and its content, including the data or business processes of the enterprise.
- **Local law where data are stored**—Local laws apply to the CSPs storage centers. The country that houses the storage center also imposes the law. This is very important, especially regarding privacy. Privacy regulations that the enterprise is bound to may not be applicable in the country where the data are housed, thus potentially compromising the stored data.
- **Local law where data are processed**—In some countries local laws do not merely apply to where data are stored, but are also applicable to the location where data are processed.

Larger CSPs will take legal matters into account when deploying their data centers and storage centers to specifically avoid legal issues. It is important, however, to have the CSPs storage regions backed up by an independent organization so it can be assured that data or business processes are in the best possible hands.

The goal is to find the CSP that best serves the standards and business needs of the enterprise. A transition to the cloud will be much easier when a CSP has the same industry certifications as the enterprise. This also goes for the business needs of the enterprise: If the business is changing rapidly, the enterprise will want to choose a CSP that can change quickly as well and that is agile. If, for example, computing requirements change daily, the enterprise will want a CSP that can accommodate those changes rapidly and fluently. A CSP that can only change processor power every two weeks while also being required to fill in four different forms may not be the best-suited choice for the enterprise. One must keep in mind that CSPs offer services for a large variety of disparate customers. And while some deployment and service models will leave room for minor adjustments, this puts the CSP at risk because it forces it to go outside its known area of expertise. The rule of thumb is that the better an offered service complies with the business needs, the more a CSP will be compliant with the enterprise.

An effective vendor management process with goals and objectives can help ensure the following:
- Vendor selection and management strategy is consistent with enterprise goals.
- Effective cooperation and governance models are in place.
- Service, quality, cost and business goals are clear.
- All parties perform as agreed.
- Vendor risk is assessed and properly addressed.
- Vendor relationships are working effectively, as measured according to service objectives

Stakeholder Responsibilities and Viewpoints

Many stakeholders are involved in the vendor management process. As explained in chapter 1. Introduction, establishing and managing good vendor relations does not involve IT or the business process owners solely, but also many other stakeholders within the enterprise. The participation of the legal function is crucial in helping to define requirements and writing contracts. Compliance and audit should be consulted when reviewing service agreements and vendor compliance assessments. Risk management and business continuity should analyze vendor-related risk, etc.

The RACI (responsible, accountable, consulted and informed) chart in **figure 26** is a high-level representation of the various stakeholders in vendor management.

Figure 26—Vendor Management RACI Chart

Stakeholders	Contractual Relationship Life Cycle			
	Setup	Contract	Operations	Transition-out
C-level executives	A	A	A	A
Business process owners	R	R	I	R
Procurement	R	R	I	R
Legal	R	R	C	C
Chief risk officer	C	C	R	R
Compliance and audit	C	C	C	C
IT	R	R	R	R
Security	R	C	R	C
HR	C	C	C	C

Life Cycle of the Contractual Relationship

The life cycle of the contractual vendor relationship can be divided into four phases, as shown in **figure 27**. Each phase typically consists of a number of activities.

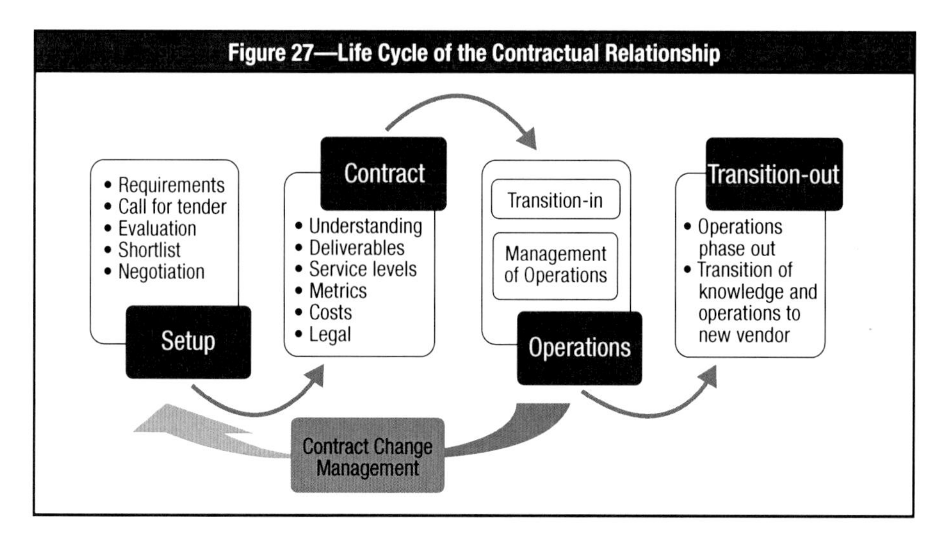

C-level Executives

The C-level executive who is accountable for the vendor management process depends on the scale of outsourcing. When the scope and impact of products and services are limited to IT, the CIO is often accountable for the execution of the contractual vendor relationship life cycle phases. When the scope also includes the business, the CIO provides oversight and coordination of all responsible parties and supports other C-level executives in the decision-making processes. The CIO provides oversight to ensure that:

- Contract objectives do not conflict among vendors, products and services.
- Contracts comply with internal policies.
- Contracts are managed properly.
- SLAs are met.

For involvement of large-scale vendor services covering or impacting the entire enterprise, however, the CEO and CFO are more closely involved and take on the accountability for the activities described previously for the CIO.

The CFO is responsible for the enterprise budget, after it is approved by the board; therefore, the CFO should approve the budget that is available for IT vendors. The CFO should be at least consulted for the following responsibilities:

- Drafting the call for tender to validate budget specifications
- Negotiating with multiple vendors
- Agreeing on the final vendor contract

The CEO and CFO should be informed of vendor performance during the operations phase.

Business Process Owners

The goal of contracting IT products or services is to enable the enterprise to achieve its goals. Therefore, business process owners should be actively involved in the vendor management life cycle in addition to IT and procurement. Some key responsibilities of business process owners include:

- Providing business requirements
- Consolidating stakeholder requirements and making sure they are validated and approved by all parties
- Evaluating potential vendors using specific business knowledge and expertise
- Providing input for contract specifications (e.g., service levels and transition-out specifications)

Procurement

Many responsibilities within the vendor management life cycle belong to the procurement function. Involving sourcing professionals from the beginning of the process enables them to support business units, IT and compliance during the selection and contract management processes. Following are some key responsibilities of procurement:

- Help business process owners consolidate business requirements from various stakeholders.
- Help IT consolidate IT-related requirements.
- Prepare and send out the call for tender.
- Gather vendor offers.
- Facilitate the vendor evaluation process.
- Facilitate vendor negotiations.
- Facilitate contract management activities.

Legal

To effectively mitigate vendor-related risk, the legal function should be involved throughout the entire vendor management life cycle. Assigning ownership of certain tasks creates commitment and a close collaboration with the other stakeholders. Key responsibilities of the legal function include:

- Provide input regarding service requirements, from a legal point of view.
- Consolidate legal requirements and make sure that they are validated and approved by the other stakeholders.
- Provide boilerplate standard contract language for important legal and compliance provisions.
- Evaluate potential vendors using specific legal knowledge and expertise.
- Draft the contract, taking into account and reviewing the specifications provided by other stakeholders.

Risk Function

The risk function should be consulted throughout the vendor management life cycle to obtain a complete view on risk that is related to the relationship, services or products. During the setup and contract phases, the risk function provides requirements to mitigate potential risk. Potential risk during the operations phase is addressed within the enterprise risk framework, based on information provided by the business units and the enterprise risk appetite.

If the vendor environment, business model, products or services change, the risk function initiates an assessment to identify and address new risk. After the risk assessment is completed, the enterprise identifies controls that are required to minimize newly identified risk and updates related documentation (e.g., business impact analyses, BCPs, DRPs, insurance policies and the risk framework).

Compliance and Audit

The compliance and audit functions should be consulted throughout the vendor management life cycle to ensure compliance with internal and external laws, regulations and policies. Compliance and audit participate in the vendor management life cycle by:

- Ensuring compliance of vendor candidates during the selection process
- Providing compliance requirements to be included in the contract
- Ensuring compliance of the selected vendor during operations

IT

All stakeholders share the responsibility of the vendor management process to ensure that products and services received from the vendor support enterprise goals within risk tolerances. However, the IT role is significant because its members may be more familiar with the products and services and their market availability. Key responsibilities of IT include the following:

- Identifying potential vendors, products and services
- Providing input regarding requirements for products and services from an IT point of view
- Providing input associated with baseline standard requirements and architectural limitations
- Providing expectations associated with good practices and defined architectural boundaries
- Providing information practices, such as information ownership and co-location requirements, partnership arrangements and employee checking
- Consolidating the previous inputs and ensuring that they are validated and approved by all stakeholders
- Evaluating potential vendors using specific IT knowledge and expertise
- Providing input for contract specifications (e.g., transition-in specifications, transition-out specifications and service metrics)
- Monitoring, reporting and managing vendor performance during operations

Security

Although security is often associated with the IT stakeholder, security is regarded as a separate stakeholder because it does not always reside within the IT function and can be involved with other functions as well as IT. Security is responsible for the following:

- Providing input regarding security requirements for products and services
- Providing input associated with baseline security standard requirements
- Providing information practices, such as information accessibility and required assurances of information protection
- Verifying compliance of the vendor with product, service and information security requirements

Human Resources

The HR stakeholder should be consulted throughout the vendor management life cycle to ensure compliance with the enterprise's worker statutes, local regulations, code of conduct and labor law.

Drafting Contracts and Service Level Agreements

The enterprise should detail in the vendor contract, among other factors, the following items:

- Fees
- Roles and responsibilities
- Deliverables
- Workflows
- Fallback procedures
- Penalty and reward mechanisms
- Confidentiality of information
- Intellectual property
- Transition-out procedures

Furthermore, the contract must conform to enterprise standards and legal and regulatory requirements. Given the differences among local regulations, positive opportunities as well as negative consequences of the local context should be considered.

Another portion of the contract should include the SLA, with specific service levels and metrics that have been accepted by the vendor.

The most important inputs for the contract are detailed requirements (business, IT, compliance, HR, security and legal), the vendor proposal, and the vendor risk assessment. The vendor risk assessment needs to be completed during the selection process and should include potential tangible and intangible impacts that the enterprise could experience. Potential risk related to the products or services required should also be evaluated and addressed by defining service level requirements and contingency procedures to mitigate possible vendor failure.

Enterprise rules and policies should be shared with the selected vendor to create awareness about the internal control environment and requirements for compliance. Important documents include the following:

- Enterprise strategy
- Information security policy
- Physical and environmental security policy
- Access control policy
- List of applicable laws and regulations

The vendor should provide assurance of compliance prior to initiating the operations phase. Sometimes, it is recommended to include some of these requirements in the call for tender to ensure that the vendor will have the capabilities necessary to comply, before starting negotiations.

4. Security Considerations for Cloud Computing

When considering the implementation strategies, service models and related risk that are discussed in the previous chapters, it is noteworthy that most of the risk-increasing factors affect theft and disclosure, while most of the risk-decreasing factors affect unavailability and loss. This could be interpreted as a trade-off.

The perception about cloud security (or insecurity) is still the main concern preventing some enterprises from adopting cloud computing as part of the IT landscape. This concern is based on the limited visibility into the provider's environment and the difficulty in assessing cloud-related risk. However, the benefits of using cloud computing are real, and these benefits can be better exploited by enterprises that have robust governance, policies and procedures to address the inherent risk of the cloud.

Security and privacy must be considered from the moment management considers cloud computing a possibility. Security and privacy requirements should be part of the due diligence, business case studies and vendor selection. Whenever possible, the enterprise should evaluate the provider's control environment to ensure that the enterprise's security and privacy needs will be met satisfactorily. When a close assessment is not possible, the enterprise must request proof that the provider has implemented the necessary controls to protect the enterprise's assets. Ideally this proof should demonstrate that the provider is capable of offering security controls that comply with or exceed the enterprise's needs.

Cloud computing benefits and risk are derived from the nature of the technologies used to support cloud services (virtualization, elasticity, resource-sharing, etc.). The list of potential risk factors described in chapter 2 shows that most of the risk-increasing factors impact theft and disclosure negatively, while most of the risk-decreasing factors impact unavailability and loss positively. This could be interpreted as a trade-off that must be factored in when deciding whether cloud computing is a good alternative for the enterprise.

Once the enterprise has selected the cloud service and deployment models, it is important to develop internal security management policies and procedures that can be used to protect the confidentiality, integrity, availability, authenticity and privacy of the information that will reside, travel or be processed in the cloud.

It is very important to understand that cloud security is the responsibility of all parties involved in the contract. The provider is responsible for implementing and maintaining the necessary controls to protect its clients' assets. The enterprise is responsible for ensuring that the security and privacy controls are implemented correctly, operate as intended and meet the enterprise's requirements throughout the life of the contract. The responsibility to protect information assets can be transferred to the provider, but accountability for the protection of information assets will remain with the enterprise.

Overview of Threats and Mitigating Actions

Figure 28 addresses the possible security threats that could exploit any of the risk-increasing factors that are described in the section Risk Assessment When Migrating to the Cloud in chapter 2, and provides links to the related factors for each threat. This section also maps the security threats to mitigating actions found in *COBIT 5 for Information Security*, which explains selected terminology in more detail and how to implement certain security mitigating actions within the enterprise.

With the implementation of these mitigating strategies, the impact and/or likelihood of a risk event are greatly reduced, based on controls deployed. Risk and threats still exist, but they are reduced. Specific risk assessments must be conducted periodically to evaluate the risk situation of the enterprise assets and to identify improvement opportunities.

Figure 28—Cloud Threats and Mitigating Actions Mapped to *COBIT 5 for Information Security*

Threats	Description	Risk Factors	Mitigation	Mapping to *COBIT 5 for Information Security*
Technical Threats				
A. Vulnerable access management (infrastructure and application, public cloud)	Information assets could be accessed by unauthorized entities due to faulty or vulnerable access management (AM) measures or processes. This could result from a forgery/theft of legitimate credentials or a common technical practice (e.g., administrator permissions override)	S1.D S3.F D1.B D2.C	• A contractual agreement is necessary to officially clarify who is allowed to access the enterprise's information, naming specific roles for CSP employees and external partners. • Request that the CSP to provide detailed technical specifications of its IAM system for the enterprise's CISO to review and approve. Include additional controls to ensure robustness of the CSPs IAM system. Most CSPs will not provide such details due to internal security policies, but the enterprise can request controls and benchmarks as an alternative (e.g., result of penetration testing on the CSPs IAM systems). • Use corporate IAM systems instead of CSPs IAM systems. The IAM remains the responsibility of the enterprise, so no access to assets can be granted without the knowledge of the enterprise. It requires the approval of the CSP and the establishment of a secure channel between the CSP infrastructure and the corporate IAM system.	• Appendix A. Detailed Guidance: Principles, Policies and Frameworks Enabler – A.2 Information Security Policy • Appendix F. Detailed Guidance: Services, Infrastructure and Applications Enabler – F.6 Provide User Access and Access Rights in Line With Business Requirements – F.10 Provide Monitoring and Alert Services for Security-related Events

Figure 28—Cloud Threats and Mitigating Actions Mapped to *COBIT 5 for Information Security* (cont.)

Threats	Description	Risk Factors	Mitigation	Mapping to *COBIT 5 for Information Security*
Technical Threats *(cont.)*				
B. Data visible to other tenants when resources are allocated dynamically	This refers to data that has been stored in memory space or disk space that can be recovered by other entities sharing the cloud by using forensics techniques.	S1.E	• A contractual agreement is necessary to officially clarify who is allowed to access the enterprise's information, naming specific roles for CSP employees and external partners. All controls protecting the enterprise's information assets must be clearly documented in the contract agreement or SLA. • Encrypt all sensitive assets that are being migrated to the CSP, and ensure that proper key management processes are in place. This will consume part of the allocated resources due to the encrypt/decrypt process so global performance could be affected. • Request the CSPs technical specifications and controls to ensure that the data are properly wiped when requested. • Use a private cloud deployment model (no multitenancy).	• Appendix G. Detailed Guidance: People, Skills and Competencies Enabler – G.3 Information Risk Management – G.6 Information Assessment and Testing and Compliance • Appendix F. Detailed Guidance: Services, Infrastructure and Applications Enabler – F.5 Provide Adequately Secured and Configured Systems, in Line With Security Requirements and Security Architecture – F.9 Provide Security Testing

Figure 28—Cloud Threats and Mitigating Actions Mapped to *COBIT 5 for Information Security* (cont.)

Threats	Description	Risk Factors	Mitigation	Mapping to *COBIT 5 for Information Security*
Technical Threats *(cont.)*				
C. Multitenancy visibility	Due to the nature of multitenancy, some assets (e.g., routing tables, MAC addresses, internal IP addresses, LAN traffic) could be visible to other entities in the same cloud. Malicious entities in the cloud could take advantage of the information, e.g., by utilizing shared routing tables to map the internal network topology of an enterprise, preparing the way for an internal attack.	S1.E D1.B D2.C	• Request the CSPs technical details for CISO approval and require additional controls to ensure data privacy. • A contractual agreement is necessary to officially clarify who is allowed to access the enterprise's information, naming specific roles for CSP employees and external partners. All controls protecting the enterprise's information assets must be clearly documented in the contract agreement or SLA. • Use a private cloud deployment model (no multitenancy).	• Appendix E. Detailed Guidance: Information Enabler – E.8 Information Security Review Reports • Appendix C. Detailed Guidance: Organizational Structures Enabler – C.1 Chief Information Security Officer (CISO) • Appendix F. Detailed Guidance: Services, Infrastructure and Applications Enabler – F.10 Provide Monitoring and Alert Services for Security-related Events
D. Hypervisor attacks	Hypervisors are vital for cloud virtualization. They provide the link between VMs and the underlying physical resources required to run the machines by using hypercalls (similar to system calls, but for virtualized systems). An attacker using a VM in the same cloud could fake hypercalls to inject malicious code or trigger bugs in the hypervisor. This could potentially be used to violate confidentiality or integrity of other VMs or crash the hypervisor (similar to a DDoS attack).	S1.E	• Request CSPs internal SLA for hypervisor vulnerability management, patch management and release management when new hypervisor vulnerabilities are discovered. The SLA must contain detailed specifications about vulnerability classification and actions taken according to the severity level. • Use a private cloud deployment model (no multitenancy).	• Appendix B. Detailed Guidance: Processes Enabler – B.2 Align, Plan and Organize: APO09 Manage Service Agreements • Appendix G. Detailed Guidance: People, Skills and Competencies Enabler – G.3 Information Risk Management • Appendix A. Detailed Guidance: Principles, Policies and Framework Enabler – A.2 Information Security Policy

Figure 28—Cloud Threats and Mitigating Actions Mapped to *COBIT 5 for Information Security* (cont.)

Threats	Description	Risk Factors	Mitigation	Mapping to *COBIT 5 for Information Security*
Technical Threats *(cont.)*				
E. Application attacks	Due to the nature of SaaS, the applications offered by a CSP are more broadly exposed. Because they can be the target of massive and elaborate application attacks, additional security measures (besides standard network firewalls) are required to protect them.	S3.H	• Request that the CSP implements application firewalls, antivirus, and antimalware tools. • The SLA must contain detailed specifications about vulnerability classification and actions taken according to the severity level, which must align with corporate policies and procedures for similar events.	• Appendix G. Detailed Guidance: People, Skills and Competencies Enabler – G.5 Information Security Operations – G.6 Information Assessment and Testing and Compliance • Appendix F. Detailed Guidance: Services, Infrastructure and Applications Enabler – F.10 Provide Monitoring and Alert Services for Security-related Events
F. Application compatibility	In a virtualized environment, direct access to resources is no longer possible (the hypervisor stays in the middle). This could generate issues with older and/or bespoke software that could go unnoticed until it is too late to fall back.	D3.C	• Evaluate extensively the design and requirements of application candidates to move to the cloud and/or request the CSP a test period to identify possible issues. • Require continuous communication and notification of major changes to ensure that compatibility testing is included in the change plans.	• Appendix B. Detailed Guidance: Processes Enabler – B.3 Build, Acquire and Implement: – BAI02 Manage Requirements Definition • Appendix E. Detailed Guidance: Information Enabler – E.6 Information Security Requirements • Appendix F. Detailed Guidance: Services, Infrastructure and Applications Enabler – F.3 Provide Secure Development
G. Collateral damage	The enterprise can be affected by issues involving other entities sharing the cloud. For example, DDoS attacks affecting another entity in the cloud can leave the enterprise without access to business applications (for SaaS models) or extra computing resources to handle peak loads (for IaaS models).	D1.C	• Ask the CSP to include the enterprise in its incident management process that deals with notification of collateral events. • Include contract clauses and controls to ensure that the enterprise's contracted capacity is always available and cannot be directed to other tenants without approval. • Use a private cloud deployment model (no multitenancy).	• Appendix E. Detailed Guidance: Information Enabler – E.6 Information Security Requirements • Appendix F. Detailed Guidance: Services, Infrastructure and Applications Enabler – F.8 Provide Adequate Incident Response • Appendix G. Detailed Guidance: People, Skills and Competencies Enabler – G.3 Information Risk Management

Figure 28—Cloud Threats and Mitigating Actions Mapped to *COBIT 5 for Information Security* (cont.)

Threats	Description	Risk Factors	Mitigation	Mapping to *COBIT 5 for Information Security*
Technical Threats *(cont.)*				
H. SaaS access security	Access to SaaS applications (either via browser of specific end-user clients) must be secure in order to control the exposure to attacks and protect the enterprise and its assets.	S3.K	• Use hardened web browsers and/or specific end-user client applications which include appropriate security measures (anti-malware, encryption, sandboxes, etc.). • Use secure virtual desktops or specific browser clients when connecting to cloud applications.	• Appendix F. Detailed Guidance: Services, Infrastructure and Applications Enabler – F.6 Provide User Access and Access Rights in Line With Business Requirements – F.10 Provide Monitoring and Alert Services for Security-related Events • Appendix G. Detailed Guidance: People, Skills and Competencies Enabler – G.5 Information Security Operations
I. Outdated VM security	An inactive VM could be easily overlooked and important security patches could be left unapplied. This out-of-date VM could become compromised when activated.	S1.K	• Introduce procedures within the enterprise to verify the state of software security updates prior to the activation of any VMs. • Empower the CSP to apply security patches on inactive VMs.	• Appendix A. Detailed Guidance: Principles, Policies and Framework Enabler – A.2 Information Security Policy • Appendix F. Detailed Guidance: Services, Infrastructure and Applications Enabler – F.5 Adequately Secured and Configured Systems, Aligned With Security Requirements and Security Architecture

Figure 28—Cloud Threats and Mitigating Actions Mapped to *COBIT 5 for Information Security* (cont.)

Threats	Description	Risk Factors	Mitigation	Mapping to *COBIT 5 for Information Security*
Regulatory Threats				
A. Asset ownership	Any asset (data, application or process) migrated to a CSP could be legally owned by the CSP based on contract terms. Thus, the enterprise can lose sensitive data or have it disclosed since the enterprise is no longer the sole legal owner of the asset. In the event of contract termination, the enterprise could even be subject (by contract) to pay fees to retrieve its own assets	S2.D S3.C	• Include terms in the contract with the CSP that ensure that the enterprise remains the sole legal owner of any asset migrated to the CSP. • Encrypt all sensitive assets being migrated to the CSP prior to the migration to prevent disclosure. This could affect the performance of the system.	• Appendix C. Detailed Guidance: Organizational Structures Enabler – C.5 Information Custodians/business Owners
B. Asset disposal	In the event of contract termination, to prevent disclosure of the enterprise's assets, those assets should be removed from the cloud using forensic tools (or other tools that ensure a complete wipeout).	S1.I S2.E S3.D	• Request CSPs technical specifications and controls that ensure that data are properly wiped and backup media is destroyed when requested. • Include terms in the contract that require, upon contract expiration or any event ending the contract, a mandatory data wipe carried out under the enterprise's review.	• Appendix G. Detailed Guidance: People, Skills and Competencies Enabler – G.3 Information Risk Management

Figure 28—Cloud Threats and Mitigating Actions Mapped to *COBIT 5 for Information Security* *(cont.)*

Threats	Description	Risk Factors	Mitigation	Mapping to *COBIT 5 for Information Security*
Regulatory Threats *(cont.)*				
C. Asset location	Technical information assets (data, logs, etc.) are subject to the regulations of the country where they are stored. In the cloud, an enterprise may, without notification, migrate information assets to countries where regulations are less restrictive or their transmission is prohibited (personal information in particular). Unauthorized entities that cannot have access to assets in one country may be able to obtain legal access in another country. Conversely, if assets are moved to countries with stricter regulations, the enterprise can be subject to legal actions and fines for noncompliance.	S1.D	• Request the CSPs list of infrastructure locations and verify that regulation in those locations is aligned with the enterprise's requirements. • Include terms in the service contract to restrict the moving of enterprise assets to only those areas known to be compliant with the enterprise's own regulation. • To prevent disclosure, encrypt any asset prior to migration to the CSP	• Appendix G. Detailed Guidance: People, Skills and Competencies Enabler – G.4 Information Security Architecture Development – G.6 Information Assessment and Testing and Compliance • Appendix F. Detailed Guidance: Services, Infrastructure and Applications Enabler – F.2 Provide Security Awareness

Figure 28—Cloud Threats and Mitigating Actions Mapped to *COBIT 5 for Information Security* (cont.)

Threats	Description	Risk Factors	Mitigation	Mapping to *COBIT 5 for Information Security*
Information Security Governance Threats				
A. Physical security on all premises where data/applications are stored	Physical security is required in any infrastructure. When the enterprise migrates assets to a cloud infrastructure, those assets are still subject to the corporate security policy but they can also be physically accessed by the CSPs staff, which is subject to the CSPs security policy. There could be a gap between the security measures provided by the CSP and the requirements of the enterprise.	S1.H	• Request the CSPs physical security policy and ensure that it is aligned with the enterprise's security policy. • Request the CSP to provide proof of independent security reviews or certifications reports (for example AICPA SSAE 16 SOC 2 report or ISO certification). • Include in the contract language that requires the CSP to be aligned with the enterprise's security policy and to implement necessary controls to ensure it. • Request the CSPs disaster recovery plans and ensure that it contains the necessary countermeasures to protect physical assets during and after a disaster.	• Appendix E. Detailed Guidance: Information Enabler – E.6 Information Security Requirements • Appendix A. Detailed Guidance: Principles, Policies and Frameworks Enabler – A.2 Information Security Policy

Figure 28—Cloud Threats and Mitigating Actions Mapped to *COBIT 5 for Information Security* (cont.)

Threats	Description	Risk Factors	Mitigation	Mapping to *COBIT 5 for Information Security*
Information Security Governance Threats *(cont.)*				
B. Visibility of the security measures put in place by the CSP	The cloud is similar to any infrastructure in that security measures (technology and processes) should be in place to prevent security attacks. The security measures provided by the CSP should be aligned with the requirements of the enterprise, including management of security incidents.	S1.F	• Request the CSPs detailed schemes of the technical security measures in place and determine whether they meet the requirements of the enterprise. • Request the CSP to provide proof of independent security reviews or certifications reports (for example AICSPA SSAE 16 SOC 2 report or ISO certification). • Include in the contract language that requires the CSP to provide the enterprise regular reporting on security (incident reports, IDS/IPS logs, etc.). • Request the CSPs security incident management process to be applied to the enterprise's assets and ensure that it is aligned with the enterprise's own security policy.	• Appendix E. Detailed Guidance: Information Enabler – E.6 Information Security Requirements – E.8 Information Security Review Reports – E.9 Information Security Dashboard • Appendix F. Detailed Guidance: Services, Infrastructure and Applications Enabler – F.10 Provide Monitoring and Alert Services for Security-related Events
C. Media Management	Data media support must be disposed in a secure way to avoid data leakage and disclosure. Data wipeout procedures must ensure data cannot be reproduced when data media support is designated for recycle or disposal. Controls should be in place during transportation (encryption and physical security). This should be specified in the CSP security policy.	S1.I	• Request the CSPs process and techniques in place for data media disposal and evaluate whether they meet the requirements of the enterprise. • Include in the contract language that requires the CSP to comply with the enterprise's security policy	• Appendix B. Detailed Guidance: Processes Enabler – B. 3 Build, Acquire and Implement: BAI08 Manage Knowledge

Figure 28—Cloud Threats and Mitigating Actions Mapped to *COBIT 5 for Information Security* (cont.)

Threats	Description	Risk Factors	Mitigation	Mapping to *COBIT 5 for Information Security*
Information Security Governance Threats *(cont.)*				
D. Secure SDLC	When using SaaS services, the enterprise must be sure that the applications will meet its security requirements. This will reduce the risk of disclosure	S3.E	• Request the CSPs details about the SDLC in place and ensure that the security measures introduced into the design are compliant with the requirements of the enterprise. • Request the CSP to provide proof of independent security reviews or certifications reports (for example AICSPA SSAE 16 SOC 2 report or ISO certification).	• Appendix B. Detailed Guidance: Processes Enabler – B. 3 Build, Acquire and Implement: BAI03 Manage Solutions Identification and Build • Appendix E. Detailed Guidance: Information Enabler – E.6 Information Security Requirements • Appendix F. Detailed Guidance: Services, Infrastructure and Applications Enabler – F.3 Provide Secure Development
E. Common security policy for community clouds	Community clouds share resources among different entities that belong to the same group (or community) and thereby possess a certain level of mutual "trust." This trust must be regulated by a common security policy. Otherwise, an attack on the "weakest link" of the group could place all the group's entities in danger.	D2.C	• Ensure that a global security policy specifying minimum requirements is applied to all entities sharing a community cloud. • Request the CSP to provide proof of independent security reviews or certifications reports (for example AICSPA SSAE 16 SOC 2 report or ISO certification).	• Appendix E. Detailed Guidance: Information Enabler – E.6 Information Security Requirements • Appendix 5. Detailed Guidance: Principles, Policies and Framework Enabler – E.2 Information Security Policy

Figure 28—Cloud Threats and Mitigating Actions Mapped to *COBIT 5 for Information Security* (cont.)

Threats	Description	Risk Factors	Mitigation	Mapping to *COBIT 5 for Information Security*
Information Security Governance Threats *(cont.)*				
F. Service termination issues	Currently, there is very little available in terms of tools, procedures or other offerings to facilitate data or service portability from CSP to CSP. This can make it very difficult for the enterprise to migrate from one CSP to another or to bring services back in-house. It can also result in serious business disruption or failure should the CSP go bankrupt, face legal action, or be the potential target for an acquisition (with the likelihood of sudden changes in CSP policies and any agreements in place). Another possibility is the "run on the banks" scenario, in which there is a crisis of confidence in the CSPs financial position resulting in a mass exit and withdrawal on first-come, first-served basis. If there are limits to the amount of content that can be withdrawn in a given time frame, then the enterprise might not be able to retrieve all its data. Another possibility may occur if the enterprise decides, for any reason, to end the relationship with the CSP. The complexity of the business logic and data models could make it literally impossible for the enterprise to extract its data, reconstruct the business logic and rebuild the applications.	S3.G	• Ensure by contract or SLA with the CSP an exit strategy that specifies the terms that should trigger the retrieval of the enterprise's assets in the time frame required by the enterprise. • Implement a DRP, taking into account the possibility of complete disruption of the CSP	• Appendix B. Detailed Guidance: Processes Enabler – B.2 Align, Plan and Organize: APO09 Manage Service Agreements • Appendix B. Detailed Guidance: Processes Enabler – B.4 Deliver, Service and Support: DSS04 Manage Continuity • Appendix G. Detailed Guidance: People, Skills and Competencies Enabler – G.3 Information Risk Management

Figure 28—Cloud Threats and Mitigating Actions Mapped to *COBIT 5 for Information Security (cont.)*

Threats	Description	Risk Factors	Mitigation	Mapping to *COBIT 5 for Information Security*
Information Security Governance Threats *(cont.)*				
G. Solid enterprise governance	Enterprises turn to CSPs in search of solutions that can be implemented easily and low-cost. This ease can be tempting, especially when the enterprise is facing urgent deadlines that require an urgent solution (like the expiration of application licenses, or the need of more computing capacity). This can become an issue because organizations may contract cloud applications without proper procurement and approval oversight, thus bypassing compliance with internal policies	S3.I	Ensure that internal governance controls are in place within the enterprise to involve the necessary control organizations (legal, compliance, finance, etc.) during the decision process of migrating to cloud services.	• Appendix B. Detailed Guidance: Processes Enabler – B.1 Evaluate, Direct and Monitor (EDM): EDM01 Ensure governance framework setting and maintenance – B.5 Monitor, Evaluate and Assess (MEA): MEA02 Monitor, Evaluate and Assess the System of Internal Control
H. Support for audit and forensic investigations	Security audits and forensic investigations are vital to the enterprise to evaluate the security measures of the CSP (preventive and corrective), and in some cases the CSP itself (for example, to authenticate the CSP). This raises several issues because performing these actions require to have extensive access to the CSPs infrastructure and monitoring capabilities, which are often shared with other CSPs customers. The enterprise should have the permission of the CSP to perform regular audits and to have access to forensic data without violating the contractual obligations of the CSPs to their other customers.	S1.F S1.L	• Request the CSP the right to audit as part of the contract or SLA. If this is not possible request security audit reports by trusted third parties. • Request the CSP to provide appropriate and timely support (logs, traces, hard disk images) for forensic analysis as part of the contract or SLA. If this is not possible, request to authorize trusted third parties to perform forensic analysis when necessary.	• Appendix B. Detailed Guidance: Processes Enabler – B.1 Align, Plan and Organise (APO): APO10 Manage suppliers. – B.5 Monitor, Evaluate and Assess (MEA): MEA02 Monitor, Evaluate and Assess the System of Internal Control

5. Assurance in Cloud Computing

The Merriam-Webster dictionary defines assurance as "something that inspires or tends to inspire confidence."[23] ISACA defines an assurance initiative as an "objective examination of evidence for the purpose of providing an assessment on risk management, control or governance processes for the organization."[24]

The previous chapters of this book have examined the challenges and risk inherent in cloud computing—challenges made even more daunting by the plethora of cloud computing solutions that have arisen over a short period of time. Before moving ahead with the decision to roll out a cloud service or utilize cloud computing, there is a strong need for assurance mechanisms.

What is certain is that assurance, and ultimately "confidence," in the cloud is different from a traditional outsourcing arrangement. Previously, assurance over outsourcing agreements was much better understood, because boundaries were established; frameworks, including certification and accreditation (C&A) by third parties, were defined and available from independent service providers; and assurance was usually provided by looking at historical transactions that could be aligned with the client or isolated to defined locations and facilities. With shared resourcing, multitenancy and geolocation in mind, cloud computing requires an entirely new approach to providing assurance. In the cloud, boundaries are difficult to define and isolate, and client-specific transactional information is difficult to obtain. Assurance needs to become more real-time, continuous and process-oriented vs. transactional in focus, while CSPs need to provide greater transparency to their clients regarding the movement of the clients' data. Security and assurance frameworks and certification and accreditation standards that are specific to CSPs must continue to evolve as clients seek confidence in the services of the CSPs.

What are the mechanisms available for obtaining assurance in the cloud computing space? Part of the answer depends on whether the one asking this question is a CSP or a cloud user. For example, cloud users want assurance over the accuracy of cloud usage metering and billing processes so that they are not overbilled; CSPs need assurance about the quality of the technology infrastructure to help ensure maximum availability and avoid revenue loss that may result if the cloud service offered is less than what was promised to the user.

Therefore, various perceptions on assurance within the cloud ecosystem exist. The main views are from the organizations that offer cloud services and the cloud users. Depending on the nature of the cloud deployment model, there could be other parties interested in assurance, e.g., a data protection authority (DPA) for citizens who store data in the cloud. A closer look at both perspectives is taken later in this chapter, but, first, the assurance requirements, standards and frameworks are discussed.

---

[23] *www.merriam-webster.com/dictionary/assurance*

[24] ISACA Glossary, *www.isaca.org/Pages/Glossary.aspx?tid=3880&char=A*

Assurance Requirements and Standards

The cloud creates multiple assurance challenges related to regulatory and compliance requirements. Traditionally, web-hosting companies have focused on a horizontal service layer, such as hosting ERP software. Business process outsourcing (BPO) companies focus on an industry vertical, such as processing of insurance claims or other financial transactions. A side benefit of the specialization is that the assurance requirements are limited and better defined. For example, ERP-hosting providers that support the processing of financial transactions for US public companies must design and implement internal controls to help their CSP clients comply with the US Sarbanes-Oxley Act of 2002. Companies processing claims for nonpublic insurance companies, captive insurance companies, or nonprofit insurers or health plans need to help those entities to comply with the Model Audit Rule (MAR) from the US National Association of Insurance Commissioners (NAIC).

However, for public CSPs or for private clouds of large diversified conglomerates, the assurance requirements can be quite broad and not well defined. In the case of public CSPs, there may not yet be enough critical mass to allow them to focus on a horizontal or vertical focus and meet a defined set of assurance requirements.

The same factor that makes the value proposition compelling for a public cloud (compared to traditional outsourced application-hosting services), i.e., the isolation of cloud service offerings from the end user, also leads to a lack of clear determination of what meets the assurance requirements of the cloud users. For community clouds that are built with a specific horizontal or vertical in mind, there may be better definitions regarding the assurance requirements, because the clouds are built for a predefined group of users. For example, prior to introducing IaaS products that are available through the US federal government's Apps.gov web site, CSPs must complete the C&A process at the Federal Information Security Management Act (FISMA) Moderate Impact Data security level, as administered by the US General Services Administration (GSA). When authority to operate is granted, IaaS services can be made available for purchase by government entities through the Apps.gov storefront.[25]

For a diversified conglomerate that decides to implement a private cloud, the assurance challenges across multiple businesses pertaining to the private cloud might be similar to the public CSPs.

Public clouds face a multitude of requirements and standards such as PCI, the US Sarbanes-Oxley Act, internal audits, privacy protection laws, audits from service auditors and external auditors, ISO certification, and customer audits. Third-party service providers can become inundated with a multitude of compliance efforts, such as processing requests for information from existing and potential clients that is related to information security practices, supporting client auditors that may not be satisfied with an independent service auditor's report and completing detailed checklists.

---

[25] US General Services Administration (GSA), "Cloud-based Infrastructure as a Service Comes to Government," 19 October 2010, *www.gsa.gov/portal/content/193441*

These one-time audits and compliance efforts take away valuable time from other activities and are expensive intrusions on the CSP. The solution to this problem is for CSPs to adopt a consistent suite of sound assurance and policy practices that cuts across horizontal service lines and industry verticals. Even after the decision is made to implement these practices, the challenge does not end because there are multitudes of assurance frameworks available, ranging from the very broad to the very narrow.

Assurance Frameworks

With the ever-changing environment and the number of cloud computing options, there is a need for a suitable assurance framework. Many enterprises have put time into customizing existing or creating new assurance frameworks for the cloud, but the environment is still evolving, and the effort to create a consistent and broadly accepted framework remains a work in progress. At the time that this publication is published, there is no single assurance framework that broadly meets the needs of every type of CSP and client.

The existing assurance frameworks can be classified into two broad categories:
- Existing, widely accepted frameworks customizable for the cloud (i.e., COBIT, ISO 2700x)
- Frameworks built for the cloud (i.e., CSA Cloud Control Matrix, Jericho Forum® Self-Assessment Scheme)

Figure 29 summarizes the various cloud standards, certifications and assurance frameworks available.

Figure 29—Cloud Standards, Certifications and Frameworks
AICPA/CICA Trust Services (SysTrust and WebTrust)—Intended to provide assurance that an enterprise's systems controls meet one or more of the Trust Services principles and related criteria. Areas addressed by the principles include security, online privacy, availability, confidentiality and processing integrity. SysTrust is similar to a SOC 1 report, but with predefined principles and criteria. However, these principles, while of the proper intent needed for cloud risk assurance, may lack the specificity required to be effective in a cloud environment. IT audit and assurance professionals could insert within these overarching controls specific risk control points, but the responsibility is on the user auditors to properly determine these more detailed control points. Also, effective in 2011, Trust Service reports can be issued as SOC 2 or SOC 3 reports under the SSAE standard noted previously.
AICPA Service Organization Control (SOC) Reports—A SOC report is an independent, third-party examination under the AICPA/Canadian Institute of Chartered Accountants (CICA) audit standards. Released under Statement of Standards for Attestation Engagements (SSAE) No. 16 and the International Standard on Assurance Engagements (ISAE) 3402, SOC reports replaced the previously used Statement on Auditing Standards (SAS) 70 third-party examination reports effective 15 June 2011. Under a SOC report, a CSP engages a CPA firm to perform an independent examination to provide the CSP clients and their internal and external auditor's assurance regarding the understanding and reliance on controls that support the CSP client's processes and systems. There are three SOC report forms available: • ***SOC 1*** reports apply to financial reporting processes and are most consistent with prior Statement on Auditing Standards 70 reports. • ***SOC 2*** and ***SOC 3*** reports are discussed in the previous row as AICPA Trust Services. SOC reports provide the client with an understanding of the nature and significance of the services provided and the relevant impact in identifying and assessing the risk and assurances by the CSP.

Figure 29—Cloud Standards, Certifications and Frameworks *(cont.)*
Background Intelligent Transfer Service (BITS)—The BITS Shared Assessment Program contains the Standardized Information Gathering (SIG) questionnaire and Agreed Upon Procedures (AUP). They are used primarily by financial operations evaluating the IT controls that their IT service providers have in place for security, privacy and business continuity. SIG is aligned with ISO/IEC 27002:2005, Payment Card Industry Data Security Standard (PCI DSS), COBIT and NIST and is also aligned with US Federal Financial Institutions Examination Council (FFIEC) guidance, the AICPA/CICA Privacy Framework and many other privacy regulatory guidance organizations. Like the other frameworks mentioned, BITS covers most, but not all, security elements of cloud computing, with a subset of the entire questionnaire. BITS has also mapped its control framework for CSPs. (In 2010, Shared Assessments published "Evaluating Cloud Risk for the Enterprise", a risk-based guide to evaluating cloud computing for the enterprise, *sharedassessments.org/media/pdf-EnterpriseCloud-SA.pdf*.)
Cloud Control Matrix—With a first version released by CSA in April 2010, this cloud security controls matrix is specifically designed to provide fundamental security principles to guide cloud vendors and assist prospective cloud clients in assessing the overall security risk of a CSP. The foundation of CSA's cloud control matrix is other industry-accepted security standards, regulations and controls frameworks. CSA's matrix is an amalgam of controls from Health Insurance Portability and Accountability Act (HIPAA), ISO/IEC 27001/27002, COBIT, PCI and NIST.
COBIT—Developed and maintained by ISACA, COBIT 5 provides management with a comprehensive framework for the management and governance of business-driven, IT-based projects and operations. Appendix A maps the COBIT 5 process practices to the cloud.
CSA STAR Certification—Closely related to the previous rating methodology is the recently released CSA STAR Certification (September 2013). Based on the perspective that enterprises that outsource services to CSPs have a number of concerns about the security of their data and information, achieving the STAR Certification will allow cloud providers of every size to give prospective customers a greater understanding of their levels of security controls. The CSA STAR certification is a technology-neutral certification based on a third-party independent assessment of the security of a CSP. It leverages the requirements of the above described ISO/IEC 27001:2005 management system standard with the CSA Cloud Control Matrix.
European Network and Information Security Agency (ENISA)—In November 2009, ENISA released a report titled "Cloud Computing—Benefits, Risks and Recommendations for Information Security."[26] This report defines a multitude of risk points in the cloud, covering the various delivery and deployment models. One of the most important recommendations in that report is the Information Assurance Framework, a set of assurance criteria designed to assess the risk of adopting cloud services, compare different cloud provider offers, obtain assurance from the selected cloud providers and reduce the assurance burden on cloud providers.[27] In April 2012, ENISA released a practical guide aimed at the procurement and governance of cloud services. This guide provides advice on questions to ask about the monitoring of security. The goal is to improve public sector customer understanding of the security of cloud services and the potential indicators and methods which can be used to provide appropriate transparency during service delivery.[28] More recently, in February 2013, ENISA also released a report entitled "Critical Cloud Computing." In the report they look at cloud computing from a Critical Information Infrastructure Protection (CIIP) perspective and look at a number of scenarios and threats relevant from a CIIP perspective, based on a survey of public sources on uptake of cloud computing and large cyberattacks and disruptions of cloud computing services.[29]

[26] ENISA, "Cloud Computing—Benefits, Risks and Recommendations for Information Security," 2009, *www.enisa.europa.eu/activities/risk-management/files/deliverables/cloud-computing-risk-assessment*

[27] ENISA, "Cloud Computing—Information Assurance Framework," 2009, *www.enisa.europa.eu/activities/risk-management/files/deliverables/cloud-computing-information-assurance-framework*

[28] ENISA, "Procure Secure," 2012, *www.enisa.europa.eu/activities/Resilience-and-CIIP/cloud-computing/procure-secure-a-guide-to-monitoring-of-security-service-levels-in-cloud-contracts*

[29] ENISA, "Critical Cloud Computing" 2012, *www.enisa.europa.eu/activities/Resilience-and-CIIP/cloud-computing/critical-cloud-computing*

Figure 29—Cloud Standards, Certifications and Frameworks *(cont.)*
Federal Risk and Authorization Management Program—FedRAMP is a US government-wide program that provides a framework for security assessments and authorizing cloud computing services. FedRAMP is designed for federal agency use, but can be used for joint authorizations and continuous security monitoring services for both government and commercial cloud computing systems.
Jericho Forum® Self-Assessment Scheme (SAS)—A guideline for vendors to self-assess the security aspects of their cloud offering and for prospective cloud clients to include into their requests for proposal (RFPs). Jericho Forum's Self-Assessment Scheme is based on the organization's "11 Commandments," released in 2006, which are design principles for effective security in de-perimeterized environments. This Self-Assessment Scheme is designed to assess cloud security tools, either applications or devices.
ISO 20000—This was the first international standard on IT service management, established in 2005. It allows organizations to certify their "design, transition, delivery and improvement of services that fulfil service requirements and provide value for both the customer and the service provider." The ISO certification is not a one-off exercise: Maintaining the certificate requires reviewing and monitoring the Information Security Management System (ISMS) on an ongoing basis.
ISO 2700x—This is the specification against which an enterprise's information security management system (ISMS) is evaluated and by which certification is granted. The objective of the standard is to "provide a model for establishing, implementing, operating, monitoring, reviewing, maintaining and improving an ISMS." The ISO certification is not a one-time exercise: Maintaining the certificate requires reviewing and monitoring the ISMS on an ongoing basis. More than 1,000 certificates have been issued across the world. Additionally, CSA's Cloud Security Matrix has been identified as an appropriate ISO control subset for the cloud.[30]
Leet Security Rating Methodology—The European Commission is increasing its efforts to promote a single market for cybersecurity products. One of their proposed actions is to improve the information available to the public by developing security labels or kite marks helping the consumer navigate the market. One of the first of this type of service labeling system is the rating methodology developed by Leet Security.[31] The main benefit of the labeling system is that it is flexible, allowing for different requirements by different customers (or even different requirements for each service by the same customer). This is realized by giving a label to a CSP's service that shows the security measures implemented by the CSP. This allows customers to ask for an "A" rated service or "B," or even less, depending on their needs regarding information security.
NIST SP 800-53—The NIST IT security controls standards, much like ISO and COBIT, contain a controls framework required to address cloud security. Also similar to ISO and COBIT, the NIST IT security controls standards form an unspecified subset of the entire framework. Note: The current draft of NIST SP 800-146 contains additional guidance for using and implementing the various cloud deployment models.

An industrious IT audit and assurance professional could readily translate specific risk points into control points, then into IT audit tests. The appropriate tests for the specific type of cloud service would need to be identified.

Other standards and frameworks are available and can be used; however, selecting the most appropriate assurance framework or combining portions from each framework requires careful planning and involvement of the relevant stakeholders. An external audit of the cloud service will likely be required for customers to gain comfort over

[30] Cloud Security Alliance (CSA), *cloudsecurityalliance.org/cm.html*
[31] *www.leetsecurity.com*

the effectiveness of the CSP controls. CSPs have been working diligently to provide transparency to clients regarding risk and controls to avoid onsite audits and build confidence with clients. Historically, enterprises have used a variety of assurance frameworks to assess the controls of outsourced services, including services provided by CSPs. **Figures 30** and **31** summarize some of the most common frameworks and their applicability to CSPs.

Figure 30—Common Framework CSP Applicability for Assurance Frameworks

Certification/Framework Description	Benefits	Challenges
AICPA/CICA Trust Services—An independent, third-party examination under the AICPA attestation standards. CSPs can engage a CPA firm to perform a Trust Services examination to provide the CSPs' clients assurance that the CSPs' system controls meet one or more of the Trust Services principles and related criteria (security, privacy, availability, confidentiality and processing integrity). Effective in 2011, Trust Service reports can be issued as SOC 2 or SOC 3 reports under the SSAE 16 standard.	• Independent, third-party assurance on which CSP clients may rely • Examination can be specifically targeted at area(s) relevant to the CSP client • Trust Services examination may be shared with prospective clients • May be more broadly focused than financial controls, e.g., operational processes and controls • Recognized and commonly known and accepted standard	Not specifically designed as a framework/certification model for cloud environments
AICPA SOC 1 Report—An independent, third-party examination under the AICPA/CICA attestation standards. Released under the Statement of Standards for Attestation Engagements (SSAE) No. 16 and the International Statement of Attestation Engagements (ISAE) 3402. CSPs can engage a CPA firm to perform a SOC 1 report to provide clients and their audit and assurance professionals' assurance regarding the understanding and reliance on controls that support the CSP clients' financial reporting processes and systems. SOC 1 reports are the replacement to Statement on Auditing Standards 70 reports, which are well-known and understood third-party examinations.	• Independent third-party assurance on which CSP clients and their auditors may rely • Recognized and commonly known and accepted assurance framework • Potential third-party assurance tool, but will most likely need to be supplemented with other assurance methods	• An SOC 1 examination is not designed as, nor intended to be, a marketing document; and, therefore, cannot be shared with prospective clients. • May be limited in scope to financial systems and controls. CSP clients must be very careful to evaluate the scope of the SOC 1 that may or may not be applicable to the CSP services provided. • Not specifically designed as a framework/ certification model for cloud environments

Figure 30—Common Framework CSP Applicability for Assurance Frameworks *(cont.)*		
Certification/Framework Description	**Benefits**	**Challenges**
BITS—Used by financial institutions to evaluate the IT controls that their IT service providers have in place for security, privacy and business continuity. Financial institutions can provide a Shared Assessments SIG questionnaire and may also provide a Shared Assessments AUP report.	• Contains defined and specific criteria across a broad range of control areas (e.g., privacy, security, continuity) • Recently established, but commonly known framework for financial services enterprises • Recently completed a white paper mapping the BITS criteria to relevant "delta" risk for CSPs • AUP report can provide an independent, third-party assurance on which users may rely	Specifically designed for and focused on financial services industries
Cloud Control Matrix—This cloud security controls matrix is specifically designed to provide fundamental security principles to guide cloud vendors and assist prospective cloud clients in assessing overall security risk of a CSP.	• Provides a framework that is specifically focused and targeted at cloud security controls • Based on other industry-accepted security standards, regulations and controls frameworks, including controls from HIPAA, ISO/IEC 27001/27002, COBIT, PCI and NIST • Focused on cloud security and risk and not across the broad ecosystem of cloud management risk	Not yet commonly understood or consistently accepted as a framework for cloud security, but quickly gaining momentum and awareness
COBIT 5—ISACA's comprehensive framework for the management and governance of business-driven IT-based projects and operations.	• Provides a comprehensive framework to evaluate cloud environments through the entire ecosystem (due diligence through delivery management) • Commonly understood and accepted framework • Mapped to cloud risk and controls by ISACA	Not originally created for cloud-specific risk.

Figure 30—Common Framework CSP Applicability for Assurance Frameworks *(cont.)*

Certification/Framework Description	Benefits	Challenges
CSA STAR Certification—The STAR Certification allows CSPs of every size to give prospective customers a greater understanding of their levels of security controls.	• Technology-neutral certification • Based on a third-party independent assessment of the security of a CSP. • Leverages the requirements of the ISO/IEC 27001:2005 management system standard together with the CSA Cloud Control Matrix.	• Still in a start-up phase • No proven efficiency yet
ENISA Information Assurance Framework for Cloud Computing—Defines multiple risk points in the cloud, covering the various delivery and deployment models. It is a detailed discussion regarding IT cloud risk.	• Provides broader guidance that is specifically focused and targeted on cloud risk and controls • Limited to risk, but an IT audit and assurance could readily translate specific risk points into control points, then into IT audit tests • Limited to cloud risk and not focused on controls or tests of controls	• Not yet commonly understood or consistently accepted as a framework for cloud risk • Does not provide independent, third-party assurance
FedRAMP—Provides a framework for A&A cloud computing services. FedRAMP is designed for US federal agency use, but can be used for joint authorizations and continuous security monitoring services for both government and commercial cloud computing systems.	• Specifically designed for cloud computing services • Common security risk model that provides a consistent baseline for cloud-based technologies that can be leveraged across the US federal government and commercial cloud services • Provides assurance through A&A	Currently limited to deployment within government agencies, but gaining momentum in awareness and across commercial industries
ISO 20000— This ISO standards series provides a service process framework and service process accreditation relative to the standards process.	• Can provide an independent, third-party certification on which CSP clients may rely. • Well-known published standards and evaluation criteria. ISO standards are more commonly used and accepted in Europe than in other areas of the world.	Attention needs to be paid to the scope of the ISO certification, as the cloud services may not be in scope of the acquired certification.

Figure 30—Common Framework CSP Applicability for Assurance Frameworks *(cont.)*		
Certification/Framework Description	**Benefits**	**Challenges**
ISO/IEC 27001/27002— Established by ISO. ISO 2700x standards provide a security framework and process accreditation relative to the standards process.	• Can provide an independent, third-party certification on which CSP clients may rely • Well-known published standards and evaluation criteria. ISO standards are more commonly used and accepted in Europe than in other areas of the world. • Cover most security risk points of cloud computing, but applicable controls are a subset of the entire ISO 2700x control spectrum	• Few enterprises are ISO 2700x-certified or understand the certification process • Attention needs to be paid to the scope of the ISO certification, as it could be irrelevant for the services being offered.
Jericho Forum® Self-assessment Scheme—The Jericho Forum's Self-assessment Scheme is a guideline for vendors to self-assess the security aspects of their cloud offering and for prospective cloud clients to include in their RFPs.	Provides a security framework that is specifically focused and targeted on cloud vendors to self-assess their cloud offering	• Not yet commonly accepted as a framework for cloud vendors • Focused on cloud vendors and self-assessment • Does not provide independent, third-party assurance
Leet Security Rating Methodology—A rating methodology developed by Leet Security for labeling the security measures of CSPs with a unified rating.	Provides a security rating methodology that enables the customers to have a clear and comparable view of CSPs' implemented security measures and provides the customers with a flexible way of determining the services that are the best fit for their security requirements.	• Not yet commonly accepted as a rating methodology • Not yet widely applied • No proven efficiency yet
NIST SP 800-53—Contains the controls required to address cloud security	• Provides a broad risk and security framework to evaluate cloud environments • Commonly understood guidance, highly respected standard-setting body	• Not specifically focused on unique cloud risk and standards • Does not provide independent, third-party assurance

Most products and services purchased today carry safety certifications or comply with regulations from governments or industry-standards commissions. A cloud security certification would provide assurance to clients that the CSP offering provides adequate security controls around client data. Cloud security certifications undoubtedly will make clients' transitions to the cloud more palatable; however, a cloud security certification cannot be viewed as a "be-all, end-all." Just as in PCI compliance, certifications will be only a minimum threshold—a starting point. CSPs will need to show additional security measures that they implement beyond that minimum threshold.

CSPs will undoubtedly put great effort into obtaining a primary cloud security assurance certification. Perhaps industry-specific cloud security charters needed for certain markets will follow, such as PCI, BITS, Health Information Trust Alliance (HITRUST) or other assessment packages. CSPs need to look at certification for upcoming finalized cloud security assurance standards as marketing tools—with certification comes greater market acceptance. Since the varied cloud security assurance programs typically attempt to cover the same or similar requirements, it may be that, once a CSP obtains one certification, others could readily fall into place.

CSPs that incorporate a standardized cloud assurance program will benefit from a reduction of client-requested operational CSP audits. CSPs can adopt a unified IT compliance approach to cover the multitude of requirements, provide a sound base for the various assurance frameworks and establish an effective compliance-monitoring program.

Unified IT Compliance Approach

Using cloud computing services may introduce an enterprise to more regulatory compliance issues. Data may be stored or transported in different geographic regions that are subject to differing regulations. Since there could be a litany of compliance requirements, it may assist enterprises utilizing cloud services to adopt a unified IT compliance approach to more efficiently manage the compliance landscape.

The benefits of a unified compliance approach include:
- Reduced risk through a structured risk management approach
- Improved monitoring of compliance
- Improved security
- Rationalized compliance requirements and control assessment processes
- Reduced burden of compliance monitoring and testing

Key Elements of a Unified IT Compliance Program

The unified IT compliance approach includes the major components shown in **figure 31**.

Figure 31—Unified IT Compliance Components	
Business Function	**Key Activities**
Governance	Provide executive oversight and visibility through ongoing status reporting based on key performance indicators (KPIs) and compliance activities.
Risk management	• Perform a periodic risk assessment. • Identify controls in a unified controls matrix to mitigate known risk. • Efficiently address applicable compliance requirements such as PCI, the US Sarbanes-Oxley Act, privacy/breach notifications, corporate policies and standards, and customer/business partner requirements. • Perform risk assessments of new projects and systems. • Periodically update the controls matrix to address changes/new risk.
Compliance	• Develop control testing/monitoring plans. • Perform control testing/monitoring procedures in a coordinated manner to reduce or eliminate duplication of efforts across the enterprise's compliance functions. • Monitor the status of risk mitigation activities for identified control gaps. • Provide support for external audit and certification activities to enable efficiencies.
Continuous improvement	• Identify and implement solutions to address aggregated control gaps. • Automate controls and monitoring activities where possible to drive efficiency.
Unified control processes	Control activities should be executed by CSP personnel or other third parties; this includes in-house and outsourced activities.

There could be duplicity of assurance effort between the CSP and its clients, so it is important to coordinate activities to attain maximum leverage. For example, in the financial services industry, the BITS-Fiscal Operations Report and Application to Participate (FISAP) standards were formulated to produce a single consistent standard.

Assurance for Cloud Service Providers

CSPs have incentives and are challenged to establish, monitor and demonstrate ongoing compliance with a set of controls that meets their own and their customers' business and regulatory requirements. The greater the assurance, the more confidence a client will have in the CSP, which results in increased adoption and deployment of cloud solutions in the industry. The level and type of assurance should be driven by the type of cloud service model (i.e., SaaS, IaaS or PaaS), the cloud deployment model (i.e., public, private, community or hybrid) and the users of the cloud. For example, a CSP offering a community cloud used by US federal government agencies needs to consider the US Federal Risk and Authorization Management Program (FedRAMP) security framework.

As discussed in previous chapters, CSPs need to have a strong governance process, including risk optimization, in place. In alignment with the overall direction of the enterprise, the CSP must execute the appropriate activities within the context of a control framework, balancing performance and compliance in achieving the governance objectives of value creation, risk management and resource optimization.

Although compliance is a strong driver for governance, users are also interested in understanding the CSP's checks and balances. Both users and the CSP have an interest in helping to ensure the long-term sustainability of the CSP. As governance and risk management systems are established, assurance must be obtained in those areas. CSPs need to determine the audit mandate and who will be involved in assurance, and they need to respond to assurance issues at a strategic level.

The objective of the assurance process is to provide feedback to the CSP and the users on the nature, scope and level of risk and compliance. The scope of the assurance can refer to a specific subject matter, such as hardware and software acquisition, and can refer to a specific period of time. The scope of assurance can include reference to:

- Specific criteria, such as reliability, effectiveness, efficiency, availability and confidentiality
- Subject matter, such as technical standards, guidance and practices, examples of which include COSO, BITS Shared Assessment, ISO and COBIT
- Professional working standards, guidelines and practices, such as those from ISACA, PCI, FedRAMP, the CSA Cloud Control Matrix, the American Institute of Certified Public Accountants (AICPA) or NIST. (Appendix B provides a cloud computing assurance program that can be used by a CSP or a client).

Assurance can also be provided for various perspectives or scopes within a CSP. It can be provided for specific objects or assets, such as internal control, data, patents, alliances, human resources, projects or programs. Assurance can be provided at various levels within the CSP—at the overall enterprise level or the level of a specific entity, such as a geographic area, data center or service offering. Assurance can also be provided for various functionalities, such as efficiency, effectiveness and security.

Managing enterprise risk requires setting and articulating the company's risk appetite/risk tolerance and establishing risk limits. If this has not been done, there are no guidelines linked to the enterprise's strategy to indicate how much or how little risk to take. As CSPs evolve, developing a risk appetite/risk tolerance will become an increasing area of focus for cloud users.

The assurance strategy for new CSPs should start with implementing an ERM framework. However, if the company's cloud offerings are an extension of an existing line of business (e.g., Amazon has traditionally been in the business of offering an online shopping platform; Amazon Web Services is an extension of its shopping platform into the CSP business), the CSP should revisit the key assumptions in the existing ERM framework (assuming there is one) in light of the new CSP business goals. Risk that applies to the CSP business will vary. An existing traditional web-hosting provider that is planning to expand its services into the CSP business may discover little new risk from that which is already in existence; enterprises that are not involved with computer hardware and software services, however, are likely to experience entirely new risk.

Assurance for Cloud Clients

Generally, there are two main approaches/techniques that a client may use to manage and measure a CSP's quality of service performance:

1. Use vendor management, including but not limited to vendor risk assessment, vendor due diligence, vendor-tiering based on the significance of the process outsourced, and contracting and SLAs as they apply to a CSP.
2. Apply independent assurance by either the CSP's auditors or client personnel outside of vendor management (involves understanding the scope of services provided, obtaining and evaluating third-party assurance reports, evaluating residual risk and determining whether an onsite visit to the service provider is required).

Assurance Through the Vendor Management Process

Users who outsource and intend to institute vendor governance should establish a vendor management office (VMO) or a specialized sourcing function. Effective vendor management programs include both a proactive (vendor governance) and reactive (vendor contract compliance review) strategy. Cloud computing requires continuous monitoring of compliance.

The key steps of vendor management are:

1. Determine an opportunity assessment approach, and develop a business case for new business processes/IT functions to be sourced.
2. Select the process/components for a vendor-sourced delivery model.
3. Document as-is and to-be flows of the sourced process, including interfaces and hand-offs.
4. Develop a vendor short list, conduct evaluations, carry out site visits, exercise due diligence, and make a final selection.
5. Engage in contract negotiations with selected vendor(s), and reach final acceptance, inclusive of legal, tax, security and regulatory factors and of a vendor(s) replacement/exit strategy. A response to the vendor risk assessment includes mechanisms to identify and evaluate vendor risk. Based on these factors, select the contracts to evaluate further (see appendix E for more details on contracting terms).
6. Exercise strategic vendor management based on governance requirements, audits (obtain independent assurance), risk management and its mitigation, SLA trends, penalties or credits, etc. As part of strategic vendor management, enterprises with vendor management processes conduct vendor risk assessments and rank/tier vendors into different categories, depending on the significance of the relationship and results of vendor due diligence efforts. Vendor risk assessment is based on a set of predefined risk categories, such as:
    - Nature and state of the relationship
    - Complexity of contract
    - Evidence of potential errors or manipulation
    - Taxation considerations
    - Strength of audit clause
7. Conduct day-to-day vendor management based on scope and change management, issue resolution and escalation, and communications. Use project management and earned value management (EVM) techniques to measure progress against goals.

8. Define agreed-on metrics for the vendors to measure and share.
9. Conduct periodic comprehensive reviews of vendors, covering all aspects of the relationship.

Traditional vendor due diligence may gain additional assurance by obtaining a third-party examination report (e.g., a SOC 1 audit report or an ISO certificate). However, given the wide variety of potential uses for cloud computing, and its increasing complexity and diminishing transparency, a SOC 1 report or ISO certification may not be sufficient. The challenge remains that personnel assigned to perform vendor due diligence may not possess a comprehensive understanding of the risk involved in cloud computing and may end up accepting a certification that is irrelevant in meeting the vendor acceptance criteria. It is also possible that the CSP does not have the relevant certification. In certain cases, it may not be cost-effective for the CSP to obtain a specific certification to meet the needs of a single client or a small group of clients. From an assurance perspective, it is important to have a right-to-audit clause in such contracts.

In traditional vendor management, there is a lot of focus on fiscal and legal aspects that are related to vendors vs. a focus on emerging risk related to the business, e.g., reputational risk. The vendor manages security based on its risk profile and not necessarily on the client's risk model. To compensate for any lenience toward this behavior of a CSP, it is necessary to set up SLAs with the CSP that specifically target the security risk related to cloud computing.

One of the key reasons that enterprises outsource is to provide better and faster service. Enterprises using cloud computing migrate to an IT solution that responds to new business needs in real time by removing or streamlining the CSP clients' silos. Because SLAs provide the basis against which CSP clients manage performance in the outsourcing process, the SLAs must be specific and measurable, both of which present challenges in cloud computing as compared to traditional outsourcing. While there is limited precedent in the cloud computing space to provide sufficient benchmarks for SLAs, as cloud computing matures, there will be a need for benchmark data; independent, third-party intermediaries could have a role to play in this area.

As it pertains to measurability, experience with traditional IT hosting services has indicated that monitoring the quality of service (QoS) and challenging the SLA are the responsibilities of the cloud user. In addition to application processing time, network lag in external cloud computing arrangements is a factor in computing total processing time. The end user is concerned about the total processing time, which cannot exceed certain thresholds.

A CSP client may not have a ready set of tools or human resources to track the QoS in a complex cloud computing environment. Some CSPs have one standard contract, and clients must "take it or leave it." Unless the CSP client offers many potential business opportunities, the CSP may not be willing to negotiate the standard contractual terms and conditions. This can be challenging for large enterprises, because their needs are broader and require tighter clauses to ensure that the liability aspects are shared between the CSP and the enterprise. Furthermore, the very attraction of cloud's pay-as-you-go model is diluted if the cloud user needs to provide a large upfront commitment. In the cloud computing space, there is a need for well-defined SLAs that are related to various aspects of availability, response time, scalability and cost savings. Cloud computing is dynamic, and this dynamic nature must be accounted for in any system that tries to enforce SLAs. Further, due to a lack of standardization, it can be difficult to compare SLAs across multiple CSPs.

Because of this situation, another service emerged—multivendor governance, also often referred to as Multisourcing Service Integrator (MSI). Usually an independent third party, the MSI coordinates the interaction not only between the client and its CSP, but also between different CSPs. This ensures that the interaction between multiple CSPs works as flawlessly as possible and can span all aspects of the cloud, ranging from fine-tuning SLAs and APIs to effectively gather all needed data, sanitizing it and making it available to all CSPs for easier cooperation. A large part of multivendor governance includes gathering all transactional and performance data from both the client as the CSPs, to get an overall view of all services. These results are then compared to the agreed-on SLAs and are used for capacity management and planning.

Issues at the end of a relationship should also be considered when contracting with CSPs. With SaaS, the data reside with the CSP, so it is important to understand the format in which the data will be available, if the CSP client decides to end the relationship. With PaaS, due to the lack of clarity, it is important to understand who owns the platform, process and data and to document the hand-offs as part of the contracting process. With IaaS, it is important to understand the VM technology and whether the CSP has done significant customization to the VM that would hinder it from using an alternative CSP.

Benefits of Effective Service Level Management

Well-designed SLAs and mature service level management maximize the effectiveness of the services they target and manage. This improves business performance and fosters a better relationship between the enterprise and the vendor through:

- Better alignment with business objectives
- Ability to manage services proactively
- Greater transparency of service delivery
- Lower service level management overhead
- Better relationships between the enterprise and vendor

Operational Level Agreements and Underpinning Contracts

An SLA is an agreement between the enterprise and a product or service provider. This service provider may be an external party; an internal department, e.g., the IT department for IT services; or a combination of both. When two or more (internal or external) groups collaborate for the delivery of these services, they usually define an operational level agreement (OLA) that applies to all parties involved.

The purpose of an OLA is to stipulate clearly and measurably the service provider's internal interdependent support relationships. In this way, the internal services are properly aligned to provide the intended SLA. Although OLAs are not as comprehensive and detailed as SLAs, the same two main sections that are in the SLA should be included in an OLA. The first section identifies services and performance metrics. The second section in the OLA discusses processes related to OLA governance and change management. The OLA should be tested and changed only when necessary, it is not static, and contains start and end dates.

Underpinning contracts can be considered external OLAs. An underpinning contract is an agreement between the vendor and a third party or subcontractor, whereby the third party provides services or goods to the vendor to support the vendor's delivery of the service to the enterprise. Thus the underpinning contract is a supporting document for the SLA and contains stipulations that are required to meet the agreed-on SLA targets.

The underpinning contract also contains sections about the identification of services and performance metrics and about governance and change management, like the SLA. The same pitfalls can also be taken into account. It is important for the enterprise to verify whether any stipulations exist in the original contract between the enterprise and the vendor that state that the vendor is (or is not) permitted, to outsource part of its activities to a third party. Some enterprises consider it to be too high a risk for the vendor to outsource part of its activities and, therefore, prohibit the vendor from outsourcing by so stating explicitly in the contract.

Assurance Provided by CSP Clients' Independent Auditors/Assessors

Third-party service providers are inundated with assurance requests; therefore, it is becoming increasingly common for CSPs to elect to undergo an audit/review/assessment from an independent auditor/assessor and share the report/certification. While C&A standards are evolving, the most commonly used reports are the ISAE 3402 (SOC 1), ISO/IEC 27001, and AICPA/CICA Trust Services (SOC 2 and SOC 3) reports. In September 2013, CSA launched a new certification, the CSA STAR Certification. This is a technology-neutral certification that is based on a third-party independent assessment of the security of a CSP. It leverages the requirements of the ISO/IEC 27001:2005 management system standard with the CSA Cloud Control Matrix. Regardless of the form of report or certification, users must always determine the impact that the CSP can have on them and evaluate the scope of the independent examination, including the completeness and adequacy of the performed testing and results.

6. Putting It All Together

The previous chapters in this book are intended to help enterprises understand the different areas that should be involved when deciding to move enterprise assets to the cloud. The holistic approach proposed in this book includes governance and risk management practices that should guide the cloud management life cycle (evaluation and selection of cloud services, transition to the cloud, CSP management, assurance and decommission), security practices to protect enterprise assets, and assurance practices to determine whether the cloud services in use meet enterprise goals and compliance requirements. The appendices are tools that can be used to accomplish some of the steps mentioned throughout the book.

Cloud computing requires oversight from the CSP and the client. Enterprises should consider CSPs as partners who share the responsibility of managing assets residing in the cloud, as shown in **figure 32**. The criticality of these partners will depend on the nature of the assets the enterprise is willing to move to the CSPs environment. The enterprise should remain as the sole proprietor of the assets, while the CSP will become a guardian responsible for the activities agreed on in the contract and SLAs.

Figure 32—Partnership for Cloud Computing

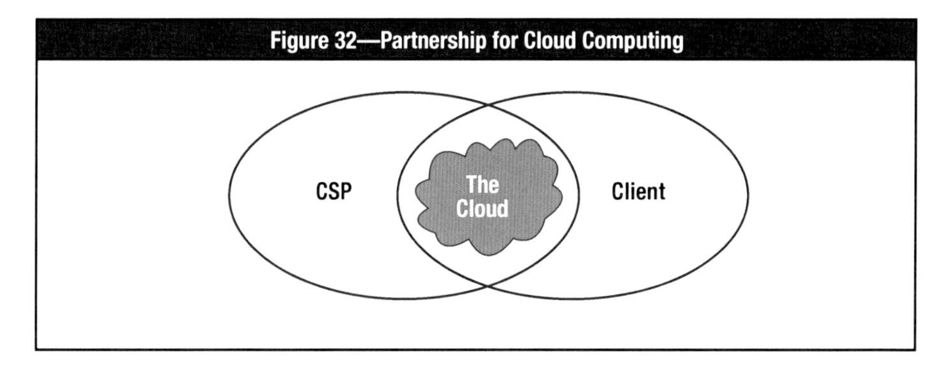

Developing partnerships is important because the enterprise is responsible for understanding the activities involved in ensuring that critical assets are protected in alignment with enterprise objectives, internal policies, applicable laws and regulations. If the enterprise decides that moving its assets to a less than optimal environment is necessary, the enterprise should seek to implement compensatory controls to reduce risk associated with identified gaps. **See figure 33**.

Figure 33—Cloud Architecture

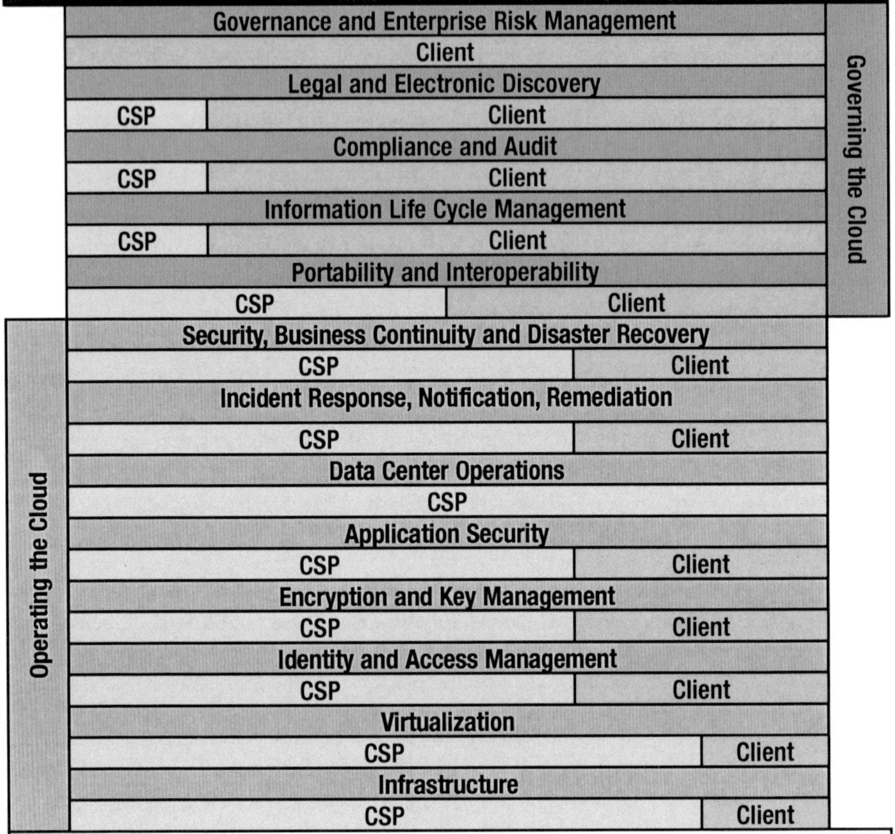

Area	Domain	CSP	Client
Governing the Cloud	Governance and Enterprise Risk Management		Client
Governing the Cloud	Legal and Electronic Discovery	CSP	Client
Governing the Cloud	Compliance and Audit	CSP	Client
Governing the Cloud	Information Life Cycle Management	CSP	Client
Governing the Cloud	Portability and Interoperability	CSP	Client
Operating the Cloud	Security, Business Continuity and Disaster Recovery	CSP	Client
Operating the Cloud	Incident Response, Notification, Remediation	CSP	Client
Operating the Cloud	Data Center Operations	CSP	
Operating the Cloud	Application Security	CSP	Client
Operating the Cloud	Encryption and Key Management	CSP	Client
Operating the Cloud	Identity and Access Management	CSP	Client
Operating the Cloud	Virtualization	CSP	Client
Operating the Cloud	Infrastructure	CSP	Client

Based on the ISACA Webinar "IT Control Objectives for Cloud Computing" Urs Fischer and Ramses Gallego, *www.slideshare.net/ramsegallego/it-control-cloudwebinaraugust11th*

The first step in developing a partnership is to understand each party's role; this understanding can help both parties define a process that will allow the enterprise to assess periodically the adequacy and effectiveness of controls protecting its assets. See **figure 34**.

Figure 34—Responsibilities by Domain

CSP	Client
Governance and Enterprise Risk Management	
	• Ensure governance framework setting and maintenance • Ensure risk optimization • Ensure resource optimization • Define and manage the cloud strategy • Communicate desire outcomes • Manage suppliers • Manage service agreements • Monitor compliance

Figure 34—Responsibilities by Domain *(cont.)*	
CSP	**Client**
Legal and Electronic Discovery	
• Meet requirements for data retention • Meet requirement for evidence protection • Provide data as needed during e-discovery or legal procedures	• Define requirements • Communicate requirements • Choose the level or package of service that best meets enterprise goals (based on budget, risk appetite and anticipated benefits delivery) • Document requirements in contracts and SLAs • Monitor compliance
Compliance and Audit	
• Establish a monitoring approach • Set performance and conformance targets • Collect and process performance and conformance data • Analyze and report performance • Ensure the implementation of corrective actions • Monitor internal controls • Review business process control effectiveness • Perform control self-assessments • Identify and report controls deficiencies • Ensure that assurance providers are independent and qualified • Plan assurance initiatives • Scope assurance initiatives • Execute assurance initiatives	• Define requirements • Communicate requirements • Choose the level or package of service that best meets enterprise goals (based on budget, risk appetite and anticipated benefits delivery) • Document requirements in contracts and SLAs • Identify changes in external compliance requirements and communicate need requirements to CSP • Optimize response to external requirements • Confirm external compliance • Obtain assurance of external compliance • Request proof of independent reviews by third parties
Information Life Cycle Management	
• Meet data management requirements • Implement adequate processes to dispose data and data storage devices • Return data to client when contracts expire	• Identify assets • Classify assets • Define requirements • Communicate requirements • Choose the level or package of service that best meets enterprise goals (based on budget, risk appetite and anticipated benefits delivery) • Document requirements in contracts and SLAs • Monitor compliance
Portability and Interoperability	
• Design, develop and implement APIs following industry accepted standards • Implement standard operational processes • Notify client of any changes	• Due diligence to select compatible CSPs • Implement a change management process to ensure compatibility and interoperability • Test new applications

Figure 34—Responsibilities by Domain *(cont.)*

CSP	Client
Security, Business Continuity and Disaster Recovery	
• Protect against malware • Manage network and connectivity security • Manage end-point security • Manage user identity and logical access • Manage physical access to IT assets • Manage sensitive documents and output devices • Monitor the infrastructure for security-related events • Define the business continuity policy, objectives and scope to meet client requirements • Maintain a continuity strategy • Develop and implement a business continuity response • Exercise, test and review the DRP • Review, maintain and improve the continuity plan • Conduct continuity plan training • Manage backup arrangements • Conduct postresumption review	• Understand environment to identify applicable laws and regulations • Define requirements • Communicate requirements • Choose the level or package of service that best meets enterprise goals (based on budget, risk appetite and anticipated benefits delivery) • Document requirements in contracts and SLAs • Assess control environment • Report results • Maintain an internal contingency plan to reduce the impact of cloud services disruptions
Incident Response, Notification and Remediation	
• Define incident and service request classification schemes • Record, classify and prioritize requests and incidents • Verify, approve and fulfill service requests • Investigate, diagnose and allocate incidents • Resolve and recover from incidents • Close service requests and incidents • Track status and produce reports	• Define requirements • Communicate requirements • Choose the level or package of service that best meets enterprise goals (based on budget, risk appetite and anticipated benefits delivery) • Document requirements in contracts and SLAs • Assess control environment • Report results • Maintain an internal incident response plan (as needed)
Data Center Operations	
• Manage problems • Manage continuity • Manage security • Perform operational procedures • Manage outsourced IT services • Monitor IT infrastructure • Manage the environment • Manage facilities • Manage network and connectivity security • Manage end-point security • Manage user identity and logical access • Manage physical access to IT assets • Monitor the infrastructure for security-related events	• Identify the geographical locations that the CSP will use to provide services • Define requirements • Communicate requirements • Choose the level or package of service that best meets enterprise goals (based on budget, risk appetite and anticipated benefits delivery) • Document requirements in contracts and SLAs • Implement a change management process to ensure that assets are relocated to acceptable locations

Figure 34—Responsibilities by Domain *(cont.)*	
CSP	**Client**
Application Security	
• Implement security controls to protect the application and the data it processes or transmits • Protect against malware • Manage user identity and logical access (client and third parties) • Align control activities embedded in business processes with enterprise objectives • Control the processing of information • Manage roles, responsibilities, access privileges and levels of authority • Manage errors and exceptions • Ensure traceability of information events and accountabilities	• Understand environment to identify applicable laws and regulations • Define requirements • Communicate requirements • Choose the level or package of service that best meets enterprise goals (based on budget, risk appetite and anticipated benefits delivery) • Document requirements in contracts and SLAs • Assess control environment • Report results • Implement a change management process to ensure changes to applications do not disable or remove security controls
Encryption and Key Management	
• Manage keys as needed • Implement the level of encryption that would meet the client's expectations	• Understand environment to identify applicable laws and regulations • Define requirements • Communicate requirements • Choose the level or package of service that best meets enterprise goals (based on budget, risk appetite and anticipated benefits delivery) • Document requirements in contracts and SLAs • Assess control environment • Report results • Implement a key management process (as needed)
Identity and Access Management	
• Manage user identity and logical access • Manage physical access to IT assets • Manage roles, responsibilities, access privileges and levels of authority • Ensure traceability of information events and accountabilities • Monitor the infrastructure for security-related events	• Define requirements • Communicate requirements • Choose the level or package of service that best meets enterprise goals (based on budget, risk appetite and anticipated benefits delivery) • Document requirements in contracts and SLAs • Implement internal processes to manage identity and access requests • Assess control environment • Report results
Virtualization	
• Establish and maintain the architecture • Manage operations • Notify client of any changes	• Define requirements • Communicate requirements • Document requirements in contracts and SLAs • Assess control environment • Report results

Figure 34—Responsibilities by Domain *(cont.)*	
CSP	**Client**
Infrastructure	
• Manage architecture • Manage operations • Manage network and connectivity security • Manage end-point security • Manage user identity and logical access • Manage physical access to IT assets • Monitor the infrastructure for security-related events	• Define requirements • Communicate requirements • Document requirements in contracts and SLAs • Assess control environment • Report results

The appendices create a toolbox for governance, security and assurance professionals. Each tool has a specific application to help the enterprise govern and manage the cloud. They can be used together or individually to assess cloud services and report the results to management. The most important thing to remember is that each tool should be customized to reflect the specific enterprise's environment and goals.

How to Use the Appendices

All of the appendices should be customized to reflect enterprise requirements. The focus of each appendix is:

- **Appendix A. COBIT 5 Governance and Management Practices—Figure 35** is intended to help the reader identify applicable practices (controls) that can be implemented to govern and manage the cloud computing services and providers. These practices are fully described in the publication *COBIT 5: Enabling Processes*. The practices have been mapped to the Cloud Security Alliance Controls Matrix version 3 to help the reader understand their relationship and eliminate duplication of controls. This table could be the foundation for a matrix that should be tailored to meet enterprise requirements for security and compliance for the cloud delivery model chosen.

  Intended Use: To create a controls framework to secure cloud services (CSP, Client or Both)

- **Appendix B. Cloud Computing Assurance Program—Figure 37** is a template that can be used as a road map for the preparation and execution of a specific assurance engagement. This assurance program template was designed in alignment with COBIT 5 and *COBIT 5 for Assurance*. It is important to note that the template is intended as an example. The template should be customized to include all enablers that are applicable to the enterprise and the cloud services in scope.

  Intended Use: To assess the controls (all enablers) and determine if they are adequate and effective (CSP, Client or Third-party Assessor)

- **Appendix C. Process Capability Assessment**—The capability assessment is an important tool for assurance initiatives. It can be part of the audit/assurance assessment or used as a metric to document the process maturity reach at a particular time. The reader should customize the stepped approach described in this appendix prior to performing an assessment.
  Intended Use: To assess the process capability, for example vendor management, operations management, change management, review of contracts and SLAs (CSA, Client, Both or Third-party Assessor)

- **Appendix D. Cloud Risk Scenarios**—Provides examples of risk scenarios specific to cloud computing. The table in **figure 41** can be used to identify and analyze risk and develop action plans as necessary. The practice of developing risk scenarios is to support risk management activities and make them more realistic and relevant to the enterprise.
  Intended Use: To describe potential risk during a risk assessment (CSP, Client, Both or Third-party Assessor)

- **Appendix E. Contractual Provisions**—Describes some of the most important contractual provisions to define and manage custody and ownership of enterprise assets. All applicable provisions should be properly documented in contracts and SLAs covering cloud services. This list of provisions is intended as guidance for nonlegal stakeholders who need to review and assess contracts and SLAs as part of audit/assurance engagements.
  Intended Use: To review contracts and SLAs as part of an assurance or process capability assessment (Client or Third-party Assessor)

- **Appendix F. Cloud Enterprise Risk Management (ERM) Governance Checklist**—Figure 42 should be used as a baseline to develop a more comprehensive list of questions the board of directors should ask as part of their oversight obligations when the enterprise has decided to adopt cloud services.
  Intended Use: To assess the governance framework or to start building one (Client or Third-party Assessor)

- **Appendix G. A Practical Approach to Measuring Cloud ROI**—Complements the section "Calculating Cloud Computing ROI" in chapter 3. Figure 43 outlines the three phases and suggested questions needed to address each step when trying to calculate the real ROI of cloud computing (tangible and intangible).
  Intended Use: To demonstrate common financial calculations that can be included in a business case to justify or reject cloud initiatives (Client or Third-party Assessor)

Page intentionally left blank

Appendix A. COBIT 5 Governance and Management Practices

Using the Cross-Reference for COBIT 5

COBIT 5 was developed as a generic governance and management framework, covering seven enablers, including Processes. This appendix adapts the COBIT 5 enabling processes to the cloud environment and identifies the process practices that are relevant to users and providers of cloud services. See **figure 35**.

All COBIT 5 process practices that are identified in this appendix have some applicability to the cloud; however, some practices are of a higher priority than others.

The following provides guidance for using this appendix:

1. For each cloud delivery model (IaaS, PaaS and SaaS), the main responsible party (client or CSP) is indicated. The two degrees of importance in responsibility, high and low, are indicated with bold text for high importance and regular text for low importance. For example, if the client's responsibility is of high importance, the cell in the table includes "**Client**". If the process is equally important to both the CSP and the client, the cell contains "Both".

   If the cell is empty, the process practice is not important for that cloud delivery model. The following example illustrates how the indicators appear.

	Low Importance	High Importance
Client	Client	**Client**
CSP	CSP	**CSP**
Client and CSP	Both	**Both**

2. A mapping is provided between each COBIT 5 process practice and the applicable Cloud Security Alliance controls from its Cloud Controls Matrix version 3.[32] The Cloud Security Alliance mission is to "promote the use of best practices for providing security assurance within Cloud Computing, and provide education on the uses of Cloud Computing to help secure all other forms of computing" and, therefore, provides excellent further guidance to COBIT's process practices for cloud computing. It is recommended that you review this mapping for future CSA updates.
3. While some controls are valid and important for all deployment models, the way they are implemented or treated depends largely on the service model chosen. This is important to keep in mind when reading the following table.

[32] CSA, *cloudsecurityalliance.org/research/ccm*

Figure 35—COBIT 5 Process Practices and the Cloud

COBIT 5 Process Practice	Main Responsibility			Mapping to CSA CCM V3 Control ID
	SaaS	PaaS	IaaS	
Evaluate, Direct and Monitor				
EDM01 Ensure Governance Framework Setting and Maintenance Analyse and articulate the requirements for the governance of enterprise IT, and put in place and maintain effective enabling structures, principles, processes and practices, with clarity of responsibilities and authority to achieve the enterprise's mission, goals and objectives.				
Process Purpose Statement Provide a consistent approach integrated and aligned with the enterprise governance approach. To ensure that IT-related decisions are made in line with the enterprise's strategies and objectives, ensure that IT-related processes are overseen effectively and transparently, compliance with legal and regulatory requirements is confirmed, and the governance requirements for board members are met.				
Key Governance Practices				
EDM01.01 Evaluate the governance system. Continually identify and engage with the enterprise's stakeholders, document an understanding of the requirements, and make a judgement on the current and future design of governance of enterprise IT.	Client	Client	Client	
EDM01.02 Direct the governance system. Inform leaders and obtain their support, buy-in and commitment. Guide the structures, processes and practices for the governance of IT in line with agreed-on governance design principles, decision-making models and authority levels. Define the information required for informed decision making.	Client	Client	Client	
EDM01.03 Monitor the governance system. Monitor the effectiveness and performance of the enterprise's governance of IT. Assess whether the governance system and implemented mechanisms (including structures, principles and processes) are operating effectively and provide appropriate oversight of IT.	Client	Client	Client	

Figure 35—COBIT 5 Process Practices and the Cloud *(cont.)*

COBIT 5 Process Practice	Main Responsibility			Mapping to CSA CCM V3 Control ID
	SaaS	PaaS	IaaS	
Evaluate, Direct and Monitor *(cont.)*				
EDM02 Ensure Benefits Delivery Optimise the value contribution to the business from the business processes, IT services and IT assets resulting from investments made by IT at acceptable costs.				
Process Purpose Statement Secure optimal value from IT-enabled initiatives, services and assets; cost-efficient delivery of solutions and services; and a reliable and accurate picture of costs and likely benefits so that business needs are supported effectively and efficiently.				
Key Governance Practices				
EDM02.01 Evaluate value optimisation. Continually evaluate the portfolio of IT-enabled investments, services and assets to determine the likelihood of achieving enterprise objectives and delivering value at a reasonable cost. Identify and make judgement on any changes in direction that need to be given to management to optimise value creation.	Client	Client	Client	
EDM02.02 Direct value optimisation. Direct value management principles and practices to enable optimal value realisation from IT-enabled investments throughout their full economic life cycle.	Client	Client	Client	
EDM02.03 Monitor value optimisation. Monitor the key goals and metrics to determine the extent to which the business is generating the expected value and benefits to the enterprise from IT-enabled investments and services. Identify significant issues and consider corrective actions.	Client	Client	Client	

Figure 35—COBIT 5 Process Practices and the Cloud *(cont.)*

COBIT 5 Process Practice	Main Responsibility			Mapping to CSA CCM V3 Control ID
	SaaS	PaaS	IaaS	
Evaluate, Direct and Monitor *(cont.)*				
EDM03 Ensure Risk Optimization Ensure that the enterprise's risk appetite and tolerance are understood, articulated and communicated, and that risk to enterprise value related to the use of IT is identified and managed.				
Process Purpose Statement Ensure that IT-related enterprise risk does not exceed risk appetite and risk tolerance, the impact of IT risk to enterprise value is identified and managed, and the potential for compliance failures is minimised.				
Key Governance Practices				
EDM03.01 Evaluate risk management. Continually examine and make judgment on the effect of risk on the current and future use of IT in the enterprise. Consider whether the enterprise's risk appetite is appropriate and that risk to enterprise value related to the use of IT is identified and managed.	**Client**	**Client**	**Client**	
EDM03.02 Direct risk management. Direct the establishment of risk management practices to provide reasonable assurance that IT risk management practices are appropriate to ensure that the actual IT risk does not exceed the board's risk appetite.	**Both**	**Both**	**Both**	BCR-10 GRM-02 GRM-11
EDM03.03 Monitor risk management. Monitor the key goals and metrics of the risk management processes and establish how deviations or problems will be identified, tracked and reported for remediation.	**Both**	**Both**	**Both**	

Figure 35—COBIT 5 Process Practices and the Cloud *(cont.)*

COBIT 5 Process Practice	Main Responsibility			Mapping to CSA CCM V3 Control ID
	SaaS	PaaS	IaaS	
Evaluate, Direct and Monitor *(cont.)*				
EDM04 Ensure Resource Optimisation Ensure that adequate and sufficient IT-related capabilities (people, process and technology) are available to support enterprise objectives effectively at optimal cost.				
Process Purpose Statement Ensure that the resource needs of the enterprise are met in the optimal manner, IT costs are optimised, and there is an increased likelihood of benefit realisation and readiness for future change.				
Key Governance Practices				
EDM04.01 Evaluate resource management. Continually examine and make judgment on the current and future need for IT-related resources, options for resourcing (including sourcing strategies), and allocation and management principles to meet the needs of the enterprise in the optimal manner.	**CSP**	**CSP** Client	**Client** CSP	
EDM04.02 Direct resource management. Ensure the adoption of resource management principles to enable optimal use of IT resources throughout their full economic life cycle.	**CSP**	**CSP** Client	**Client** CSP	
EDM04.03 Monitor resource management. Monitor the key goals and metrics of the resource management processes and establish how deviations or problems will be identified, tracked and reported for remediation.	**CSP**	**CSP** Client	**Client** CSP	

Figure 35—COBIT 5 Process Practices and the Cloud *(cont.)*

COBIT 5 Process Practice	Main Responsibility			Mapping to CSA CCM V3 Control ID
	SaaS	PaaS	IaaS	
Evaluate, Direct and Monitor *(cont.)*				
EDM05 Ensure Stakeholder Transparency Ensure that enterprise IT performance and conformance measurement and reporting are transparent, with stakeholders approving the goals and metrics and the necessary remedial actions.				
Process Purpose Statement Make sure that the communication to stakeholders is effective and timely and the basis for reporting is established to increase performance, identify areas for improvement, and confirm that IT-related objectives and strategies are in line with the enterprise's strategy.				
Key Governance Practices				
EDM05.01 Evaluate stakeholder reporting requirements. Continually examine and make judgement on the current and future requirements for stakeholder communication and reporting, including both mandatory reporting requirements (e.g., regulatory) and communication to other stakeholders. Establish the principles for communication.	**CSP** Client	**Both**	**Both**	
EDM05.02 Direct stakeholder communication and reporting. Ensure the establishment of effective stakeholder communication and reporting, including mechanisms for ensuring the quality and completeness of information, oversight of mandatory reporting, and creating a communication strategy for stakeholders.	**CSP** Client	**Both**	**Both**	DCS-04
EDM05.03 Monitor stakeholder communication. Monitor the effectiveness of stakeholder communication. Assess mechanisms for ensuring accuracy, reliability and effectiveness, and ascertain whether the requirements of different stakeholders are met.	**CSP** Client	**Both**	**Both**	

Figure 35—COBIT 5 Process Practices and the Cloud *(cont.)*

COBIT 5 Process Practice	Main Responsibility			Mapping to CSA CCM V3 Control ID
	SaaS	PaaS	IaaS	
Align, Plan and Organise				
APO01 Manage the IT Management Framework Clarify and maintain the governance of enterprise IT mission and vision. Implement and maintain mechanisms and authorities to manage information and the use of IT in the enterprise in support of governance objectives in line with guiding principles and policies.				
Process Purpose Statement Provide a consistent management approach to enable the enterprise governance requirements to be met, covering management processes, organisational structures, roles and responsibilities, reliable and repeatable activities, and skills and competencies.				
Key Management Practices				
APO01.01 Define the organisational structure. Establish an internal and extended organisational structure that reflects business needs and IT priorities. Put in place the required management structures (e.g.,committees) that enable management decision making to take place in the most effective and efficient manner.	**Both**	**Both**	**Client** CSP	BCR-11 SEF-01
APO01.02 Establish roles and responsibilities. Establish, agree on and communicate roles and responsibilities of IT personnel, as well as other stakeholders with responsibilities for enterprise IT, that clearly reflect overall business needs and IT objectives and relevant personnel's authority, responsibilities and accountability.	**Both**	**Both**	**Client** CSP	BCR-11 CCC-01 DCS-04 GRM-05 HRS-04,07,08,11,12 IAM-02 SEF-01

Figure 35—COBIT 5 Process Practices and the Cloud *(cont.)*

COBIT 5 Process Practice	Main Responsibility			Mapping to CSA CCM V3 Control ID
	SaaS	PaaS	IaaS	
Align, Plan and Organise *(cont.)*				
APO01.03 Maintain the enablers of the management system. Maintain the enablers of the management system and control environment for enterprise IT, and ensure that they are integrated and aligned with the enterprise's governance and management philosophy and operating style. These enablers include the clear communication of expectations/requirements. The management system should encourage cross-divisional co-operation and teamwork, promote compliance and continuous improvement, and handle process deviations (including failure).	**Both**	**Both**	**Both**	BCR-10,11 GRM-02,03,05, 06,07,11 HRS-03,07 through 12 IAM-01,02,04 through 12 IVS-04 IPY-02 MOS-01,03 through 08, 11,12,13, 15 through 20 SEF-01 through 04 STA-03,07 TVM-01,02,03
APO01.04 Communicate management objectives and direction. Communicate awareness and understanding of IT objectives and direction to appropriate stakeholders and users throughout the enterprise.	**Both**	**Both**	**Both**	BCR-11 GRM-03,05,06 MOS-02
APO01.05 Optimise the placement of the IT function. Position the IT capability in the overall organisational structure to reflect an enterprise model relevant to the importance of IT within the enterprise, specifically its criticality to enterprise strategy and the level of operational dependence on IT. The reporting line of the CIO should be commensurate with the importance of IT within the enterprise.	**CSP**	**CSP** Client	**User** CSP	BCR-11
APO01.06 Define information (data) and system ownership. Define and maintain responsibilities for ownership of information (data) and information systems. Ensure that owners make decisions about classifying information and systems and protecting them in line with this classification.	**Both**	**Both**	**Both**	BCR-11 CCC-01 DSI-01 through 04, 06,07,08 DCS-01 EKM-01,04 GRM-01 IPY-02

Figure 35—COBIT 5 Process Practices and the Cloud *(cont.)*

COBIT 5 Process Practice	Main Responsibility			Mapping to CSA CCM V3 Control ID
	SaaS	PaaS	IaaS	
Align, Plan and Organise *(cont.)*				
APO01.07 Manage continual improvement of processes. Assess, plan and execute the continual improvement of processes and their maturity to ensure that they are capable of delivering against enterprise, governance, management and control objectives. Consider COBIT process implementation guidance, emerging standards, compliance requirements, automation opportunities, and the feedback of process users, the process team and other stakeholders. Update the process and consider impacts on process enablers.	**CSP**	**CSP** Client	**Client** CSP	BCR-11
APO01.08 Maintain compliance with policies and procedures. Put in place procedures to maintain compliance with and performance measurement of policies and other enablers of the control framework, and enforce the consequences of non-compliance or inadequate performance. Track trends and performance and consider these in the future design and improvement of the control framework.	**Both**	**Both**	**Both**	BCR-11 GRM-03,05,07 HRS-01,06 through 12 IAM-01,02,04 through 12 IVS-03,04,05,12 IPY-03 through 05 MOS-02 through 08 SEF-01 STA-09

Figure 35—COBIT 5 Process Practices and the Cloud *(cont.)*

COBIT 5 Process Practice	Main Responsibility			Mapping to CSA CCM V3 Control ID
	SaaS	PaaS	IaaS	
Align, Plan and Organise *(cont.)*				
APO02 Manage Strategy Provide a holistic view of the current business and IT environment, the future direction, and the initiatives required to migrate to the desired future environment. Leverage enterprise architecture building blocks and components, including externally provided services and related capabilities to enable nimble, reliable and efficient response to strategic objectives.				
Process Purpose Statement Align strategic IT plans with business objectives. Clearly communicate the objectives and associated accountabilities so they are understood by all, with the IT strategic options identified, structured and integrated with the business plans.				
Key Management Practices				
APO02.01 Understand enterprise direction. Consider the current enterprise environment and business processes, as well as the enterprise strategy and future objectives. Consider also the external environment of the enterprise (industry drivers, relevant regulations, basis for competition).	**Both**	**Both**	**Both**	
APO02.02 Assess the current environment, capabilities and performance. Assess the performance of current internal business and IT capabilities and external IT services, and develop an understanding of the enterprise architecture in relation to IT. Identify issues currently being experienced and develop recommendations in areas that could benefit from improvement. Consider service provider differentiators and options and the financial impact and potential costs and benefits of using external services.	**CSP** Client	**CSP** Client	**Both**	
APO02.03 Define the target IT capabilities. Define the target business and IT capabilities and required IT services. This should be based on the understanding of the enterprise environment and requirements; the assessment of the current business process and IT environment and issues; and consideration of reference standards, good practices and validated emerging technologies or innovation proposals.	**CSP**	**CSP** Client	**Both**	

Figure 35—COBIT 5 Process Practices and the Cloud *(cont.)*

COBIT 5 Process Practice	Main Responsibility			Mapping to CSA CCM V3 Control ID
	SaaS	PaaS	IaaS	
Align, Plan and Organise *(cont.)*				
APO02.04 Conduct a gap analysis. Identify the gaps between the current and target environments and consider the alignment of assets (the capabilities that support services) with business outcomes to optimise investment in and utilisation of the internal and external asset base. Consider the critical success factors to support strategy execution.	**Both**	**Both**	**Both**	
APO02.05 Define the strategic plan and road map. Create a strategic plan that defines, in co-operation with relevant stakeholders, how IT-related goals will contribute to the enterprise's strategic goals. Include how IT will support IT-enabled investment programmes, business processes, IT services and IT assets. Direct IT to define the initiatives that will be required to close the gaps, the sourcing strategy and the measurements to be used to monitor achievement of goals, then prioritise the initiatives and combine them in a high-level road map.	**Both**	**Both**	**Both**	IPY-03 through 05
APO02.06 Communicate the IT strategy and direction. Create awareness and understanding of the business and IT objectives and direction, as captured in the IT strategy, through communication to appropriate stakeholders and users throughout the enterprise.	**Both**	**Both**	**Both**	

Figure 35—COBIT 5 Process Practices and the Cloud *(cont.)*

COBIT 5 Process Practice	Main Responsibility			Mapping to CSA CCM V3 Control ID
	SaaS	PaaS	IaaS	
Align, Plan and Organise *(cont.)*				
APO03 Manage Enterprise Architecture Establish a common architecture consisting of business process, information, data, application and technology architecture layers for effectively and efficiently realising enterprise and IT strategies by creating key models and practices that describe the baseline and target architectures. Define requirements for taxonomy, standards, guidelines, procedures, templates and tools, and provide a linkage for these components. Improve alignment, increase agility, improve quality of information and generate potential cost savings through initiatives such as reuse of building block components.				
Process Purpose Statement Represent the different building blocks that make up the enterprise and their inter-relationships as well as the principles guiding their design and evolution over time, enabling a standard, responsive and efficient delivery and strategic objectives.				
Key Management Practices				
APO03.01 Develop the enterprise architecture vision. The architecture vision provides a first-cut, high-level description of the baseline and target architectures, covering the business, information, data and application and technology domains. The architecture vision provides the sponsor with a key tool to sell the benefits of the proposed capability to stakeholders within the enterprise. The architecture vision describes how the new capability will meet enterprise goals and strategic objectives and address stakeholder concerns when implemented.	**CSP**	**CSP** Client	**Both**	DSI-02 IVS-06, 08 through 10 IPY-02 through 05 MOS-10 STA-03
APO03.02 Define reference architecture. The reference architecture describes the current and target architectures for the business, information, data, application and technology domains.	**CSP**	**CSP** Client	**Both**	DSI-01 through 04, 07 DCS-01, 04 GRM-01 IVS-06, 08 through 10 IPY-03 through 05 MOS-10 STA-03

Figure 35—COBIT 5 Process Practices and the Cloud *(cont.)*

COBIT 5 Process Practice	Main Responsibility			Mapping to CSA CCM V3 Control ID
	SaaS	PaaS	IaaS	
Align, Plan and Organise *(cont.)*				
APO03.03 Select opportunities and solutions. Rationalise the gaps between baseline and target architectures, taking both business and technical perspectives, and logically group them into project work packages. Integrate the project with any related IT-enabled investment programmes to ensure that the architectural initiatives are aligned with and enable these initiatives as part of overall enterprise change. Make this a collaborative effort with key enterprise stakeholders from business and IT to assess the enterprise's transformation readiness, and identify opportunities, solutions and all implementation constraints.	**CSP**	**CSP** Client	**Both**	
APO03.04 Define architecture implementation. Create a viable implementation and migration plan in alignment with the programme and project portfolios. Ensure that the plan is closely co-ordinated to ensure that value is delivered and the required resources are available to complete the necessary work.	**CSP**	**CSP** Client	**Both**	IVS-10
APO03.05 Provide enterprise architecture services. The provision of enterprise architecture services within the enterprise includes guidance to and monitoring of implementation projects, formalising ways of working through architecture contracts, and measuring and communicating architecture's value-add and compliance monitoring.	**CSP**	**CSP** Client	**Both**	

Figure 35—COBIT 5 Process Practices and the Cloud *(cont.)*

COBIT 5 Process Practice	Main Responsibility			Mapping to CSA CCM V3 Control ID
	SaaS	PaaS	IaaS	
Align, Plan and Organise *(cont.)*				
APO04 Manage Innovation Maintain an awareness of information technology and related service trends, identify innovation opportunities, and plan how to benefit from innovation in relation to business needs. Analyse what opportunities for business innovation or improvement can be created by emerging technologies, services or IT-enabled business innovation, as well as through existing established technologies and by business and IT process innovation. Influence strategic planning and enterprise architecture decisions.				
Process Purpose Statement Achieve competitive advantage, business innovation, and improved operational effectiveness and efficiency by exploiting information technology developments.				
Key Management Practices				
APO04.01 Create an environment conducive to innovation. Create an environment that is conducive to innovation, considering issues such as culture, reward, collaboration, technology forums, and mechanisms to promote and capture employee ideas.	**CSP**	**CSP** Client	**Both**	
APO04.02 Maintain an understanding of the enterprise environment. Work with relevant stakeholders to understand their challenges. Maintain an adequate understanding of enterprise strategy and the competitive environment or other constraints so that opportunities enabled by new technologies can be identified.	**CSP**	**CSP** Client	**Both**	IVS-05 IPY-03 through 05 MOS-02, 10
APO04.03 Monitor and scan the technology environment. Perform systematic monitoring and scanning of the enterprise's external environment to identify emerging technologies that have the potential to create value (e.g.,by realising the enterprise strategy, optimising costs, avoiding obsolescence, and better enabling enterprise and IT processes). Monitor the marketplace, competitive landscape, industry sectors, and legal and regulatory trends to be able to analyse emerging technologies or innovation ideas in the enterprise context.	**CSP**	**CSP** Client	**Both**	IVS-05

Figure 35—COBIT 5 Process Practices and the Cloud *(cont.)*

COBIT 5 Process Practice	Main Responsibility			Mapping to CSA CCM V3 Control ID
	SaaS	PaaS	IaaS	
Align, Plan and Organise *(cont.)*				
APO04.04 Assess the potential of emerging technologies and innovation ideas. Analyse identified emerging technologies and/or other IT innovation suggestions. Work with stakeholders to validate assumptions on the potential of new technologies and innovation.	**CSP**	**CSP** Client	**Both**	IVS-05
APO04.05 Recommend appropriate further initiatives. Evaluate and monitor the results of proof-of-concept initiatives and, if favourable, generate recommendations for further initiatives and gain stakeholder support.	**CSP**	**CSP** Client	**Both**	
APO04.06 Monitor the implementation and use of innovation. Monitor the implementation and use of emerging technologies and innovations during integration, adoption and for the full economic life cycle to ensure that the promised benefits are realised and to identify lessons learned.	**CSP**	**CSP** Client	**Both**	

Figure 35—COBIT 5 Process Practices and the Cloud *(cont.)*

COBIT 5 Process Practice	Main Responsibility			Mapping to CSA CCM V3 Control ID
	SaaS	PaaS	IaaS	
Align, Plan and Organise *(cont.)*				
APO05 Manage Portfolio Execute the strategic direction set for investments in line with the enterprise architecture vision and the desired characteristics of the investment and related services portfolios, and consider the different categories of investments and the resources and funding constraints. Evaluate, prioritise and balance programmes and services, managing demand within resource and funding constraints, based on their alignment with strategic objectives, enterprise worth and risk. Move selected programmes into the active services portfolio for execution. Monitor the performance of the overall portfolio of services and programmes, proposing adjustments as necessary in response to programme and service performance or changing enterprise priorities.				
Process Purpose Statement Optimise the performance of the overall portfolio of programmes in response to programme and service performance and changing enterprise priorities and demands.				
Key Management Practices				
APO05.01 Establish the target investment mix. Review and ensure clarity of the enterprise and IT strategies and current services. Define an appropriate investment mix based on cost, alignment with strategy, and financial measures such as cost and expected ROI over the full economic life cycle, degree of risk, and type of benefit, for the programmes in the portfolio. Adjust the enterprise and IT strategies where necessary.	Both	Both	Both	
APO05.02 Determine the availability and sources of funds. Determine potential sources of funds, different funding options and the implications of the funding source on the investment return expectations.	Both	Both	Both	
APO05.03 Evaluate and select programmes to fund. Based on the overall investment portfolio mix requirements, evaluate and prioritise programme business cases, and decide on investment proposals. Allocate funds and initiate programmes.	Both	Both	Both	
APO05.04 Monitor, optimise and report on investment portfolio performance. On a regular basis, monitor and optimise the performance of the investment portfolio and individual programmes throughout the entire investment life cycle.	Both	Both	Both	
APO05.05 Maintain portfolios. Maintain portfolios of investment programmes and projects, IT services and IT assets.	Both	Both	Both	
APO05.06 Manage benefits achievement. Monitor the benefits of providing and maintaining appropriate IT services and capabilities, based on the agreed-on and current business case.	Both	Both	Both	

Figure 35—COBIT 5 Process Practices and the Cloud *(cont.)*

COBIT 5 Process Practice	Main Responsibility			Mapping to CSA CCM V3 Control ID
	SaaS	PaaS	IaaS	
Align, Plan and Organise *(cont.)*				
APO06 Manage Budget and Costs Manage the IT-related financial activities in both the business and IT functions, covering budget, cost and benefit management, and prioritisation of spending through the use of formal budgeting practices and a fair and equitable system of allocating costs to the enterprise. Consult stakeholders to identify and control the total costs and benefits within the context of the IT strategic and tactical plans, and initiate corrective action where needed.				
Process Purpose Statement Foster partnership between IT and enterprise stakeholders to enable the effective and efficient use of IT-related resources and provide transparency and accountability of the cost and business value of solutions and services. Enable the enterprise to make informed decisions regarding the use of IT solutions and services.				
Key Management Practices				
APO06.01 Manage finance and accounting. Establish and maintain a method to account for all IT-related costs, investments and depreciation as an integral part of the enterprise financial systems and chart of accounts to manage the investments and costs of IT. Capture and allocate actual costs, analyse variances between forecasts and actual costs, and report using the enterprise's financial measurement systems.	Both	Both	Both	
APO06.02 Prioritise resource allocation. Implement a decision-making process to prioritise the allocation of resources and rules for discretionary investments by individual business units. Include the potential use of external service providers and consider the buy, develop and rent options.	Both	Both	Both	

Figure 35—COBIT 5 Process Practices and the Cloud *(cont.)*

COBIT 5 Process Practice	Main Responsibility			Mapping to CSA CCM V3 Control ID
	SaaS	PaaS	IaaS	
Align, Plan and Organise *(cont.)*				
AP006.03 Create and maintain budgets. Prepare a budget reflecting the investment priorities supporting strategic objectives based on the portfolio of IT-enabled programmes and IT services.	Both	Both	Both	
AP006.04 Model and allocate costs. Establish and use an IT costing model based on the service definition, ensuring that allocation of costs for services is identifiable, measurable and predictable, to encourage the responsible use of resources including those provided by service providers. Regularly review and benchmark the appropriateness of the cost/chargeback model to maintain its relevance and appropriateness to the evolving business and IT activities.	Both	Both	Both	
AP006.05 Manage costs. Implement a cost management process comparing actual costs to budgets. Costs should be monitored and reported and, in case of deviations, identified in a timely manner and their impact on enterprise processes and services assessed.	Both	Both	Both	

Figure 35—COBIT 5 Process Practices and the Cloud *(cont.)*

COBIT 5 Process Practice	Main Responsibility			Mapping to CSA CCM V3 Control ID
	SaaS	PaaS	IaaS	
Align, Plan and Organise *(cont.)*				
APO07 Manage Human Resources Provide a structured approach to ensure optimal structuring, placement, decision rights and skills of human resources. This includes communicating the defined roles and responsibilities, learning and growth plans, and performance expectations, supported with competent and motivated people.				
Process Purpose Statement Optimise human resources capabilities to meet enterprise objectives.				
Key Management Practices				
APO07.01 Maintain adequate and appropriate staffing. Evaluate staffing requirements on a regular basis or upon major changes to the enterprise or operational or IT environments to ensure that the enterprise has sufficient human resources to support enterprise goals and objectives. Staffing includes both internal and external resources.	**CSP**	**CSP** Client	**Both**	BCR-11 HRS-02
APO07.02 Identify key IT personnel. Identify key IT personnel while minimising reliance on a single individual performing a critical job function through knowledge capture (documentation), knowledge sharing, succession planning and staff backup.	**CSP**	**CSP** Client	**Both**	
APO07.03 Maintain the skills and competencies of personnel. Define and manage the skills and competencies required of personnel. Regularly verify that personnel have the competencies to fulfil their roles on the basis of their education, training and/or experience, and verify that these competencies are being maintained, using qualification and certification programmes where appropriate. Provide employees with ongoing learning and opportunities to maintain their knowledge, skills and competencies at a level required to achieve enterprise goals.	**CSP**	**CSP** Client	**Both**	BCR-11 HRS-05, 10 through 12 MOS-01 SEF-03
APO07.04 Evaluate employee job performance. Perform timely performance evaluations on a regular basis against individual objectives derived from the enterprise's goals, established standards, specific job responsibilities, and the skills and competency framework. Employees should receive coaching on performance and conduct whenever appropriate.	**CSP**	**CSP** Client	**Both**	GRM-07
APO07.05 Plan and track the usage of IT and business human resources. Understand and track the current and future demand for business and IT human resources with responsibilities for enterprise IT. Identify shortfalls and provide input into sourcing plans, enterprise and IT recruitment processes sourcing plans, and business and IT recruitment processes.	**CSP**	**CSP** Client	**Both**	HRS-02, 04

Figure 35—COBIT 5 Process Practices and the Cloud *(cont.)*

COBIT 5 Process Practice	Main Responsibility			Mapping to CSA CCM V3 Control ID
	SaaS	PaaS	IaaS	
Align, Plan and Organise *(cont.)*				
APO07.06 Manage contract staff. Ensure that consultants and contract personnel who support the enterprise with IT skills know and comply with the organisation's policies and meet agreed-on contractual requirements.	**Both**	**Both**	**Both**	CCC-02 HRS-01 through 04, 07, 08, 10 through 12 IAM-07, 09 MOS-01 SEF-03

Figure 35—COBIT 5 Process Practices and the Cloud *(cont.)*

COBIT 5 Process Practice	Main Responsibility			Mapping to CSA CCM V3 Control ID
	SaaS	PaaS	IaaS	
Align, Plan and Organise *(cont.)*				
APO08 Manage Relationships Manage the relationship between the business and IT in a formalised and transparent way that ensures a focus on achieving a common and shared goal of successful enterprise outcomes in support of strategic goals and within the constraint of budgets and risk tolerance. Base the relationship on mutual trust, using open and understandable terms and common language and a willingness to take ownership and accountability for key decisions.				
Process Purpose Statement Create improved outcomes, increased confidence, trust in IT and effective use of resources.				
Key Management Practices				
APO08.01 Understand business expectations. Understand current business issues and objectives and business expectations for IT. Ensure that requirements are understood, managed and communicated, and their status agreed on and approved.	**Both**	**Both**	**Both**	DSI-01, 03, 04 DCS-01 IPY-02
APO08.02 Identify opportunities, risk and constraints for IT to enhance the business. Identify potential opportunities for IT to be an enabler of enhanced enterprise performance.	**Both**	**Both**	**Both**	
APO08.03 Manage the business relationship. Manage the relationship with customers (business representatives). Ensure that relationship roles and responsibilities are defined and assigned, and communication is facilitated.	**Both**	**Both**	**Both**	
APO08.04 Co-ordinate and communicate. Work with stakeholders and co-ordinate the end-to-end delivery of IT services and solutions provided to the business.	**Both**	**Both**	**Both**	IVS-02
APO08.05 Provide input to the continual improvement of services. Continually improve and evolve IT-enabled services and service delivery to the enterprise to align with changing enterprise and technology requirements.	**Both**	**Both**	**Both**	

Figure 35—COBIT 5 Process Practices and the Cloud *(cont.)*

COBIT 5 Process Practice	Main Responsibility			Mapping to CSA CCM V3 Control ID
	SaaS	PaaS	IaaS	
Align, Plan and Organise *(cont.)*				
APO09 Manage Service Agreements Align IT-enabled services and service levels with enterprise needs and expectations, including identification, specification, design, publishing, agreement, and monitoring of IT services, service levels and performance indicators.				
Process Purpose Statement Ensure that IT services and service levels meet current and future enterprise needs.				
Key Management Practices				
APO09.01 Identify IT services. Analyse business requirements and the way in which IT-enabled services and service levels support business processes. Discuss and agree on potential services and service levels with the business, and compare them with the current service portfolio to identify new or changed services or service level options.	**Both**	**Both**	**Both**	AIS-02, 04 DSI-02
APO09.02 Catalogue IT-enabled services. Define and maintain one or more service catalogues for relevant target groups. Publish and maintain live IT-enabled services in the service catalogues.	**Both**	**Both**	**Both**	AIS-02, 04
APO09.03 Define and prepare service agreements. Define and prepare service agreements based on the options in the service catalogues. Include internal operational agreements.	**Both**	**Both**	**Both**	AIS-01, 02, 04 BCR-10, 11 CCC-02 DSI-01, 04 DCS-01, 05 EKM-02 HRS-03, 07, 08 IPY-02 through 05 MOS-01 STA-02, 03, 05, 07, 08
APO09.04 Monitor and report service levels. Monitor service levels, report on achievements and identify trends. Provide the appropriate management information to aid performance management.	**Both**	**Both**	**Both**	CCC-02 STA-02, 07
APO09.05 Review service agreements and contracts. Conduct periodic reviews of the service agreements and revise when needed.	**Both**	**Both**	**Both**	STA-05, 07

Figure 35—COBIT 5 Process Practices and the Cloud *(cont.)*

COBIT 5 Process Practice	Main Responsibility			Mapping to CSA CCM V3 Control ID
	SaaS	PaaS	IaaS	
Align, Plan and Organise ***(cont.)***				
APO10 Manage Suppliers Manage IT-related services provided by all types of suppliers to meet enterprise requirements, including the selection of suppliers, management of relationships, management of contracts, and reviewing and monitoring of supplier performance for effectiveness and compliance.				
Process Purpose Statement Minimise the risk associated with non-performing suppliers and ensure competitive pricing.				
Key Management Practices				
APO10.01 Identify and evaluate supplier relationships and contracts. Identify suppliers and associated contracts and categorise them into type, significance and criticality. Establish supplier and contract evaluation criteria and evaluate the overall portfolio of existing and alternative suppliers and contracts.	**Both**	**Both**	**Both**	CCC-02 HRS-03 STA-01, 07
APO10.02 Select suppliers. Select suppliers according to a fair and formal practice to ensure a viable best fit based on specified requirements. Requirements should be optimised with input from potential suppliers.	**Both**	**Both**	**Both**	STA-01
APO10.03 Manage supplier relationships and contracts. Formalise and manage the supplier relationship for each supplier. Manage, maintain and monitor contracts and service delivery. Ensure that new or changed contracts conform to enterprise standards and legal and regulatory requirements. Deal with contractual disputes.	**Both**	**Both**	**Both**	STA-01, 07
APO10.04 Manage supplier risk. Identify and manage risk relating to suppliers' ability to continually provide secure, efficient and effective service delivery.	**Both**	**Both**	**Both**	CCC-02 DCS-05 HRS-07, 08 IAM-07 through 09 MOS-01 STA-01, 02, 06, 07
APO10.05 Monitor supplier performance and compliance. Periodically review the overall performance of suppliers, compliance to contract requirements, and value for money, and address identified issues.	**Both**	**Both**	**Both**	CCC-02 DCS-05 STA-01, 02, 06, 09

Figure 35—COBIT 5 Process Practices and the Cloud *(cont.)*

COBIT 5 Process Practice	Main Responsibility			Mapping to CSA CCM V3 Control ID
	SaaS	PaaS	IaaS	
Align, Plan and Organise *(cont.)*				
APO11 Manage Quality Define and communicate quality requirements in all processes, procedures and the related enterprise outcomes, including controls, ongoing monitoring, and the use of proven practices and standards in continuous improvement and efficiency efforts.				
Process Purpose Statement Ensure consistent delivery of solutions and services to meet the quality requirements of the enterprise and satisfy stakeholder needs.				
Key Management Practices				
APO11.01 Establish a quality management system (QMS). Establish and maintain a QMS that provides a standard, formal and continuous approach to quality management for information, enabling technology and business processes that are aligned with business requirements and enterprise quality management.	**Both**	**Both**	**Both**	CCC-02, 03 STA-01
APO11.02 Define and manage quality standards, practices and procedures. Identify and maintain requirements, standards, procedures and practices for key processes to guide the enterprise in meeting the intent of the agreed-on QMS. This should be in line with the IT control framework requirements. Consider certification for key processes, organisational units, products or services.	**Both**	**Both**	**Both**	CCC-02, 03 STA-01
APO11.03 Focus quality management on customers. Focus quality management on customers by determining their requirements and ensuring alignment with the quality management practices.	**CSP**	**CSP**	**CSP**	STA-01
APO11.04 Perform quality monitoring, control and reviews. Monitor the quality of processes and services on an ongoing basis as defined by the QMS. Define, plan and implement measurements to monitor customer satisfaction with quality as well as the value the QMS provides. The information gathered should be used by the process owners to improve quality.	**CSP** Client	**Both**	**Both**	CCC-02, 03 STA-01
APO11.05 Integrate quality management into solutions for development and service delivery. Incorporate relevant quality management practices into the definition, monitoring, reporting and ongoing management of solutions development and service offerings.	**CSP**	**Both**	**Both**	CCC-02, 03 STA-01
APO11.06 Maintain continuous improvement. Maintain and regularly communicate an overall quality plan that promotes continuous improvement. This should include the need for, and benefits of, continuous improvement. Collect and analyse data about the QMS, and improve its effectiveness. Correct non-conformities to prevent recurrence. Promote a culture of quality and continual improvement.	**CSP**	**CSP** Client	**Both**	STA-01

Figure 35—COBIT 5 Process Practices and the Cloud *(cont.)*

COBIT 5 Process Practice	Main Responsibility			Mapping to CSA CCM V3 Control ID
	SaaS	PaaS	IaaS	
Align, Plan and Organise *(cont.)*				
APO12 Manage Risk Continually identify, assess and reduce IT-related risk within levels of tolerance set by enterprise executive management.				
Process Purpose Statement Integrate the management of IT-related enterprise risk with overall ERM, and balance the costs and benefits of managing IT-related enterprise risk.				
Key Management Practices				
APO12.01 Collect data. Identify and collect relevant data to enable effective IT-related risk identification, analysis and reporting.	Both	Both	Both	AAC-03 GRM-02, 08 through 11
APO12.02 Analyse risk. Develop useful information to support risk decisions that take into account the business relevance of risk factors.	Both	Both	Both	AAC-03 GRM-02, 08 through 11
APO12.03 Maintain a risk profile. Maintain an inventory of known risk and risk attributes (including expected frequency, potential impact and responses) and of related resources, capabilities and current control activities.	Both	Both	Both	AAC-03 BCR-10 GRM-02, 08 through 11
APO12.04 Articulate risk. Provide information on the current state of IT-related exposures and opportunities in a timely manner to all required stakeholders for appropriate response.	Both	Both	Both	AAC-01, 02 GRM-02, 08 through 12
APO12.05 Define a risk management action portfolio. Manage opportunities to reduce risk to an acceptable level as a portfolio.	Both	Both	Both	AAC-01, 02 GRM-08 through 12
APO12.06 Respond to risk. Respond in a timely manner with effective measures to limit the magnitude of loss from IT-related events.	Both	Both	Both	AAC-01 GRM-08 through 12

Figure 35—COBIT 5 Process Practices and the Cloud *(cont.)*

COBIT 5 Process Practice	Main Responsibility			Mapping to CSA CCM V3 Control ID
	SaaS	PaaS	IaaS	
Align, Plan and Organise *(cont.)*				
APO13 Manage Security Define, operate and monitor a system for information security management.				
Process Purpose Statement Keep the impact and occurrence of information security incidents within the enterprise's risk appetite levels.				
Key Management Practices				
APO13.01 Establish and maintain an information security management system (ISMS). Establish and maintain an ISMS that provides a standard, formal and continuous approach to security management for information, enabling secure technology and business processes that are aligned with business requirements and enterprise security management.	**CSP** Client	**Both**	**Both**	AIS-01, 02, 04 CCC-04 DSI-01, 03, 04, 05, 07, 08 DCS-02, 03, 05 through 09 EKM-01 through 03 GRM-01, 04, 05, 06, 08, 09 HRS-01, 03, 06 through 12 IAM-01 through 04, 13 IVS-01, 02, 03, 06, through 12 MOS-01 through 08, 10 through 13, 15 through 20 SEF-02 through 04 TVM-01 through 03

Figure 35—COBIT 5 Process Practices and the Cloud *(cont.)*				
COBIT 5 Process Practice	**Main Responsibility**			**Mapping to CSA CCM V3 Control ID**
	SaaS	**PaaS**	**IaaS**	
Align, Plan and Organise *(cont.)*				
APO13.02 Define and manage an information security risk treatment plan. Maintain an information security plan that describes how information security risk is to be managed and aligned with the enterprise strategy and enterprise architecture. Ensure that recommendations for implementing security improvements are based on approved business cases and implemented as an integral part of services and solutions development, then operated as an integral part of business operation.	**Both**	**Both**	**Both**	DSI-03, 05 DCS-07 through 09 EKM-02 GRM-01, 04 through 06 HRS-06, 09 IAM-01, 02, 04 through 13 IVS-01, 03, 06 through 12 MOS-02 through 08, 10 through 13, 15 through 20 SEF-02 through 04 TVM-01 through 03
APO13.03 Monitor and review the ISMS. Maintain and regularly communicate the need for, and benefits of, continuous information security improvement. Collect and analyse data about the ISMS, and improve the effectiveness of the ISMS. Correct non-conformities to prevent recurrence. Promote a culture of security and continual improvement.	**CSP** Client	**Both**	**Both**	DSI-05, 07 GRM-04, 05, 08, 09 HRS-07, 08, 10 through 12 MOS-02 through 06

Figure 35—COBIT 5 Process Practices and the Cloud *(cont.)*

COBIT 5 Process Practice	Main Responsibility			Mapping to CSA CCM V3 Control ID
	SaaS	PaaS	IaaS	
Build, Acquire and Implement				
BAI01 Manage Programmes and Projects Manage all programmes and projects from the investment portfolio in alignment with enterprise strategy and in a co-ordinated way. Initiate, plan, control, and execute programmes and projects, and close with a post-implementation review.				
Process Purpose Statement Realise business benefits and reduce the risk of unexpected delays, costs and value erosion by improving communications to and involvement of business and end users, ensuring the value and quality of project deliverables and maximising their contribution to the investment and services portfolio.				
Key Management Practices				
BAI01.01 Maintain a standard approach for programme and project management. Maintain a standard approach for programme and project management that enables governance and management review and decision making and delivery management activities focussed on achieving value and goals (requirements, risk, costs, schedule, quality) for the business in a consistent manner.	Both	Both	Both	DSI-06
BAI01.02 Initiate a programme. Initiate a programme to confirm the expected benefits and obtain authorisation to proceed. This includes agreeing on programme sponsorship, confirming the programme mandate through approval of the conceptual business case, appointing programme board or committee members, producing the programme brief, reviewing and updating the business case, developing a benefits realisation plan, and obtaining approval from sponsors to proceed.	Both	Both	Both	
BAI01.03 Manage stakeholder engagement. Manage stakeholder engagement to ensure an active exchange of accurate, consistent and timely information that reaches all relevant stakeholders. This includes planning, identifying and engaging stakeholders and managing their expectations.	Both	Both	Both	
BAI01.04 Develop and maintain the programme plan. Formulate a programme to lay the initial groundwork and to position it for successful execution by formalising the scope of the work to be accomplished and identifying the deliverables that will satisfy its goals and deliver value. Maintain and update the programme plan and business case throughout the full economic life cycle of the programme, ensuring alignment with strategic objectives and reflecting the current status and updated insights gained to date.	Both	Both	Both	

Figure 35—COBIT 5 Process Practices and the Cloud *(cont.)*

COBIT 5 Process Practice	Main Responsibility			Mapping to CSA CCM V3 Control ID
	SaaS	PaaS	IaaS	
Build, Acquire and Implement *(cont.)*				
BAI01.05 Launch and execute the programme. Launch and execute the programme to acquire and direct the resources needed to accomplish the goals and benefits of the programme as defined in the programme plan. In accordance with stage-gate or release review criteria, prepare for stage-gate, iteration or release reviews to report on the progress of the programme and to be able to make the case for funding up to the following stage-gate or release review.	Both	Both	Both	
BAI01.06 Monitor, control and report on the programme outcomes. Monitor and control programme (solution delivery) and enterprise (value/outcome) performance against plan throughout the full economic life cycle of the investment. Report this performance to the programme steering committee and the sponsors.	Both	Both	Both	
BAI01.07 Start up and initiate projects within a programme. Define and document the nature and scope of the project to confirm and develop amongst stakeholders a common understanding of project scope and how it relates to other projects within the overall IT-enabled investment programme. The definition should be formally approved by the programme and project sponsors.	Both	Both	Both	
BAI01.08 Plan projects. Establish and maintain a formal, approved integrated project plan (covering business and IT resources) to guide project execution and control throughout the life of the project. The scope of projects should be clearly defined and tied to building or enhancing business capability.	Both	Both	Both	
BAI01.09 Manage programme and project quality. Prepare and execute a quality management plan, processes and practices, aligned with the QMS that describes the programme and project quality approach and how it will be implemented. The plan should be formally reviewed and agreed on by all parties concerned and then incorporated into the integrated programme and project plans.	Both	Both	Both	
BAI01.10 Manage programme and project risk. Eliminate or minimise specific risk associated with programmes and projects through a systematic process of planning, identifying, analysing, responding to and monitoring and controlling the areas or events that have the potential to cause unwanted change. Risk faced by programme and project management should be established and centrally recorded.	Both	Both	Both	

Figure 35—COBIT 5 Process Practices and the Cloud *(cont.)*

COBIT 5 Process Practice	Main Responsibility			Mapping to CSA CCM V3 Control ID
	SaaS	PaaS	IaaS	
Build, Acquire and Implement *(cont.)*				
BAI01.11 Monitor and control projects. Measure project performance against key project performance criteria such as schedule, quality, cost and risk. Identify any deviations from the expected. Assess the impact of deviations on the project and overall programme, and report results to key stakeholders.	Both	Both	Both	
BAI01.12 Manage project resources and work packages. Manage project work packages by placing formal requirements on authorising and accepting work packages, and assigning and co-ordinating appropriate business and IT resources.	Both	Both	Both	
BAI01.13 Close a project or iteration. At the end of each project, release or iteration, require the project stakeholders to ascertain whether the project, release or iteration delivered the planned results and value. Identify and communicate any outstanding activities required to achieve the planned results of the project and the benefits of the programme, and identify and document lessons learned for use on future projects, releases, iterations and programmes.	Both	Both	Both	
BAI01.14 Close a programme. Remove the programme from the active investment portfolio when there is agreement that the desired value has been achieved or when it is clear it will not be achieved within the value criteria set for the programme.	Both	Both	Both	

Figure 35—COBIT 5 Process Practices and the Cloud *(cont.)*

COBIT 5 Process Practice	Main Responsibility			Mapping to CSA CCM V3 Control ID
	SaaS	PaaS	IaaS	
Build, Acquire and Implement *(cont.)*				
BAI02 Manage Requirements Definition Identify solutions and analyse requirements before acquisition or creation to ensure that they are in line with enterprise strategic requirements covering business processes, applications, information/data, infrastructure and services. Co-ordinate with affected stakeholders the review of feasible options including relative costs and benefits, risk analysis, and approval of requirements and proposed solutions.				
Process Purpose Statement Create feasible optimal solutions that meet enterprise needs while minimising risk.				
Key Management Practices				
BAI02.01 Define and maintain business functional and technical requirements. Based on the business case, identify, prioritise, specify and agree on business information, functional, technical and control requirements covering the scope/understanding of all initiatives required to achieve the expected outcomes of the proposed IT-enabled business solution.	**CSP** Client	**Both**	**Both**	AIS-02 GRM-01 IVS-06, 07 IPY-03 through 05 MOS-08, 10 STA-03
BAI02.02 Perform a feasibility study and formulate alternative solutions. Perform a feasibility study of potential alternative solutions, assess their viability and select the preferred option. If appropriate, implement the selected option as a pilot to determine possible improvements.	**CSP** Client	**Both**	**Both**	AIS-02
BAI02.03 Manage requirements risk. Identify, document, prioritise and mitigate functional, technical and information processing-related risk associated with the enterprise requirements and proposed solution.	**Both**	**Both**	**Both**	AIS-02 DCS-04 GRM-01
BAI02.04 Obtain approval of requirements and solutions. Co-ordinate feedback from affected stakeholders and, at predetermined key stages, obtain business sponsor or product owner approval and sign-off on functional and technical requirements, feasibility studies, risk analyses and recommended solutions.	**Both**	**Both**	**Both**	AIS-02 CCC-01, 03 DCS-04 GRM-01 IPY-01, 03 through 05 MOS-08 STA-03

Figure 35—COBIT 5 Process Practices and the Cloud *(cont.)*

COBIT 5 Process Practice	Main Responsibility			Mapping to CSA CCM V3 Control ID
	SaaS	PaaS	IaaS	
Build, Acquire and Implement *(cont.)*				
BAI03 Manage Solutions Identification and Build Establish and maintain identified solutions in line with enterprise requirements covering design, development, procurement/sourcing and partnering with suppliers/vendors. Manage configuration, test preparation, testing, requirements management and maintenance of business processes, applications, information/data, infrastructure and services.				
Process Purpose Statement Establish timely and cost-effective solutions capable of supporting enterprise strategic and operational objectives.				
Key Management Practices				
BAI03.01 Design high-level solutions. Develop and document high-level designs using agreed-on and appropriate phased or rapid agile development techniques. Ensure alignment with the IT strategy and enterprise architecture. Reassess and update the designs when significant issues occur during detailed design or building phases or as the solution evolves. Ensure that stakeholders actively participate in the design and approve each version.	**CSP**	**CSP** Client	**Both**	AIS-01 IPY-01
BAI03.02 Design detailed solution components. Develop, document and elaborate detailed designs progressively using agreed-on and appropriate phased or rapid agile development techniques, addressing all components (business processes and related automated and manual controls, supporting IT applications, infrastructure services and technology products, and partners/suppliers). Ensure that the detailed design includes internal and external SLAs and OLAs.	**CSP**	**CSP** Client	**Both**	AIS-01 IVS-06, 07 IPY-01
BAI03.03 Develop solution components. Develop solution components progressively in accordance with detailed designs following development methods and documentation standards, quality assurance (QA) requirements, and approval standards. Ensure that all control requirements in the business processes, supporting IT applications and infrastructure services, services and technology products, and partners/suppliers are addressed.	**CSP**	**Both**	**Both**	AIS-01 IVS-06, 07 IPY-01 MOS-10
BAI03.04 Procure solution components. Procure solution components based on the acquisition plan in accordance with requirements and detailed designs, architecture principles and standards, and the enterprise's overall procurement and contract procedures, QA requirements, and approval standards. Ensure that all legal and contractual requirements are identified and addressed by the supplier.	**CSP**	**CSP** Client	**Both**	IVS-06, 07 IPY-01 MOS-10

Figure 35—COBIT 5 Process Practices and the Cloud *(cont.)*

COBIT 5 Process Practice	Main Responsibility			Mapping to CSA CCM V3 Control ID
	SaaS	PaaS	IaaS	
Build, Acquire and Implement *(cont.)*				
BAI03.05 Build solutions. Install and configure solutions and integrate with business process activities. Implement control, security and auditability measures during configuration, and during integration of hardware and infrastructural software, to protect resources and ensure availability and data integrity. Update the services catalogue to reflect the new solutions.	CSP	Both	Both	AIS-01 IVS-03, 06, 07 IPY-01
BAI03.06 Perform quality assurance (QA). Develop, resource and execute a QA plan aligned with the QMS to obtain the quality specified in the requirements definition and the enterprise's quality policies and procedures.	CSP	Both	Both	CCC-03
BAI03.07 Prepare for solution testing. Establish a test plan and required environments to test the individual and integrated solution components, including the business processes and supporting services, applications and infrastructure.	CSP	Both	Both	DSI-06 MOS-07
BAI03.08 Execute solution testing. Execute testing continually during development, including control testing, in accordance with the defined test plan and development practices in the appropriate environment. Engage business process owners and end users in the test team. Identify, log and prioritise errors and issues identified during testing.	CSP	Both	Both	CCC-03 MOS-07
BAI03.09 Manage changes to requirements. Track the status of individual requirements (including all rejected requirements) throughout the project life cycle and manage the approval of changes to requirements.	Both	Both	Both	DCS-04
BAI03.10 Maintain solutions. Develop and execute a plan for the maintenance of solution and infrastructure components. Include periodic reviews against business needs and operational requirements.	CSP	Both	Both	BCR-07 MOS-10
BAI03.11 Define IT services and maintain the service portfolio. Define and agree on new or changed IT services and service level options. Document new or changed service definitions and service level options to be updated in the services portfolio.	Both	Both	Both	

Figure 35—COBIT 5 Process Practices and the Cloud *(cont.)*

COBIT 5 Process Practice	Main Responsibility			Mapping to CSA CCM V3 Control ID
	SaaS	PaaS	IaaS	
Build, Acquire and Implement *(cont.)*				
BAI04 Manage Availability and Capacity Balance current and future needs for availability, performance and capacity with cost-effective service provision. Include assessment of current capabilities, forecasting of future needs based on business requirements, analysis of business impacts, and assessment of risk to plan and implement actions to meet the identified requirements.				
Process Purpose Statement Maintain service availability, efficient management of resources, and optimisation of system performance through prediction of future performance and capacity requirements.				
Key Management Practices				
BAI04.01 Assess current availability, performance and capacity and create a baseline. Assess availability, performance and capacity of services and resources to ensure that cost-justifiable capacity and performance are available to support business needs and deliver against SLAs. Create availability, performance and capacity baselines for future comparison.	**Both**	**Both**	**Both**	IVS-04
BAI04.02 Assess business impact. Identify important services to the enterprise, map services and resources to business processes, and identify business dependencies. Ensure that the impact of unavailable resources is fully agreed-on and accepted by the customer. Ensure that, for vital business functions, the SLA availability requirements can be satisfied.	**Both**	**Both**	**Both**	
BAI04.03 Plan for new or changed service requirements. Plan and prioritise availability, performance and capacity implications of changing business needs and service requirements.	**Both**	**Both**	**Both**	BCR-07
BAI04.04 Monitor and review availability and capacity. Monitor, measure, analyse, report and review availability, performance and capacity. Identify deviations from established baselines. Review trend analysis reports identifying any significant issues and variances, initiating actions where necessary, and ensuring that all outstanding issues are followed up.	**Both**	**Both**	**Both**	BCR-07 IVS-04
BAI04.05 Investigate and address availability, performance and capacity issues. Address deviations by investigating and resolving identified availability, performance and capacity issues.	**Both**	**Both**	**Both**	IVS-04

Figure 35—COBIT 5 Process Practices and the Cloud *(cont.)*

COBIT 5 Process Practice	Main Responsibility			Mapping to CSA CCM V3 Control ID
	SaaS	PaaS	IaaS	
Build, Acquire and Implement *(cont.)*				
BAI05 Manage Organizational Change Enablement Maximize the likelihood of successfully implementing sustainable enterprise wide organizational change quickly and with reduced risk, covering the complete life cycle of the change and all affected stakeholders in the business and IT.				
Process Purpose Statement Prepare and commit stakeholders for business change and reduce the risk of failure.				
Key Management Practices				
BAI05.01 Establish the desire to change. Understand the scope and impact of the envisioned change and stakeholder readiness/willingness to change. Identify actions to motivate stakeholders to accept and want to make the change work successfully.	**CSP** Client	**CSP** Client	**Both**	
BAI05.02 Form an effective implementation team. Establish an effective implementation team by assembling appropriate members, creating trust, and establishing common goals and effectiveness measures.	**CSP** Client	**CSP** Client	**Both**	
BAI05.03 Communicate desired vision. Communicate the desired vision for the change in the language of those affected by it. The communication should be made by senior management and include the rationale for, and benefits of, the change; the impacts of not making the change; and the vision, the road map and the involvement required of the various stakeholders.	**CSP**	**CSP** Client	**Both**	
BAI05.04 Empower role players and identify short-term wins. Empower those with implementation roles by ensuring that accountabilities are assigned, providing training, and aligning organisational structures and HR processes. Identify and communicate short-term wins that can be realised and are important from a change enablement perspective.	**CSP**	**CSP** Client	**Both**	

Figure 35—COBIT 5 Process Practices and the Cloud *(cont.)*

COBIT 5 Process Practice	Main Responsibility			Mapping to CSA CCM V3 Control ID
	SaaS	PaaS	IaaS	
Build, Acquire and Implement *(cont.)*				
BAI05.05 Enable operation and use. Plan and implement all technical, operational and usage aspects such that all those who are involved in the future state environment can exercise their responsibility.	**CSP** Client	**CSP** Client	**Both**	
BAI05.06 Embed new approaches. Embed the new approaches by tracking implemented changes, assessing the effectiveness of the operation and use plan, and sustaining ongoing awareness through regular communication. Take corrective measures as appropriate, which may include enforcing compliance.	**CSP** Client	**CSP** Client	**Both**	
BAI05.07 Sustain changes. Sustain changes through effective training of new staff, ongoing communication campaigns, continued top management commitment, adoption monitoring and sharing of lessons learned across the enterprise.	**CSP** Client	**CSP** Client	**Both**	

Figure 35—COBIT 5 Process Practices and the Cloud *(cont.)*

COBIT 5 Process Practice	Main Responsibility			Mapping to CSA CCM V3 Control ID
	SaaS	PaaS	IaaS	
Build, Acquire and Implement *(cont.)*				
BAI06 Manage Changes Manage all changes in a controlled manner, including standard changes and emergency maintenance relating to business processes, applications and infrastructure. This includes change standards and procedures, impact assessment, prioritization and authorization, emergency changes, tracking, reporting, closure and documentation.				
Process Purpose Statement Enable fast and reliable delivery of change to the business and mitigation of the risk of negatively impacting the stability or integrity of the changed environment.				
Key Management Practices				
BAI06.01 Evaluate, prioritise and authorise change requests. Evaluate all requests for change to determine the impact on business processes and IT services, and to assess whether change will adversely affect the operational environment and introduce unacceptable risk. Ensure that changes are logged, prioritised, categorised, assessed, authorised, planned and scheduled.	**CSP**	**Both**	**Both**	BCR-09 CCC-01, 04, 05 DCS-04 EKM-02 GRM-01 IVS-02 MOS-09, 15 TVM-02
BAI06.02 Manage emergency changes. Carefully manage emergency changes to minimise further incidents and make sure the change is controlled and takes place securely. Verify that emergency changes are appropriately assessed and authorised after the change.	**CSP**	**Both**	**Both**	CCC-05 IVS-02 MOS-09, 15 TVM-02
BAI06.03 Track and report change status. Maintain a tracking and reporting system to document rejected changes, communicate the status of approved and in-process changes, and complete changes. Make certain that approved changes are implemented as planned.	**CSP**	**Both**	**Both**	CCC-05 DSI-02 MOS-15 TVM-02
BAI06.04 Close and document the changes. Whenever changes are implemented, update accordingly the solution and user documentation and the procedures affected by the change.	**CSP**	**Both**	**Both**	CCC-05 MOS-09, 15 TVM-02

Figure 35—COBIT 5 Process Practices and the Cloud *(cont.)*

COBIT 5 Process Practice	Main Responsibility			Mapping to CSA CCM V3 Control ID
	SaaS	PaaS	IaaS	
Build, Acquire and Implement *(cont.)*				
BAI07 Manage Change Acceptance and Transitioning Formally accept and make operational new solutions, including implementation planning, system and data conversion, acceptance testing, communication, release preparation, promotion to production of new or changed business processes and IT services, early production support, and a post-implementation review.				
Process Purpose Statement Implement solutions safely and in line with the agreed-on expectations and outcomes.				
Key Management Practices				
BAI07.01 Establish an implementation plan. Establish an implementation plan that covers system and data conversion, acceptance testing criteria, communication, training, release preparation, promotion to production, early production support, a fallback/backout plan, and a post-implementation review. Obtain approval from relevant parties.	**Both**	**Both**	**Both**	CCC-05
BAI07.02 Plan business process, system and data conversion. Prepare for business process, IT service data and infrastructure migration as part of the enterprise's development methods, including audit trails and a recovery plan should the migration fail.	**Both**	**Both**	**Both**	
BAI07.03 Plan acceptance tests. Establish a test plan based on enterprisewide standards that define roles, responsibilities, and entry and exit criteria. Ensure that the plan is approved by relevant parties.	**Both**	**Both**	**Both**	CCC-03, 05
BAI07.04 Establish a test environment. Define and establish a secure test environment representative of the planned business process and IT operations environment, performance and capacity, security, internal controls, operational practices, data quality and privacy requirements, and workloads.	**Both**	**Both**	**Both**	CCC-05 DSI-06

Figure 35—COBIT 5 Process Practices and the Cloud *(cont.)*

COBIT 5 Process Practice	Main Responsibility			Mapping to CSA CCM V3 Control ID
	SaaS	PaaS	IaaS	
Build, Acquire and Implement *(cont.)*				
BAI07.05 Perform acceptance tests. Test changes independently in accordance with the defined test plan prior to migration to the live operational environment.	**Both**	**Both**	**Both**	CCC-03, 05 STA-03
BAI07.06 Promote to production and manage releases. Promote the accepted solution to the business and operations. Where appropriate, run the solution as a pilot implementation or in parallel with the old solution for a defined period and compare behaviour and results. If significant problems occur, revert back to the original environment based on the fallback/backout plan. Manage releases of solution components.	**Both**	**Both**	**Both**	CCC-05
BAI07.07 Provide early production support. Provide early support to the users and IT operations for an agreed-on period of time to deal with issues and help stabilise the new solution.	**Both**	**Both**	**Both**	
BAI07.08 Perform a post-implementation review. Conduct a post-implementation review to confirm outcome and results, identify lessons learned, and develop an action plan. Evaluate and check the actual performance and outcomes of the new or changed service against the predicted performance and outcomes (i.e., the service expected by the user or customer).	**Both**	**Both**	**Both**	

Figure 35—COBIT 5 Process Practices and the Cloud *(cont.)*

COBIT 5 Process Practice	Main Responsibility			Mapping to CSA CCM V3 Control ID
	SaaS	PaaS	IaaS	
Build, Acquire and Implement *(cont.)*				
BAI08 Manage Knowledge Maintain the availability of relevant, current, validated and reliable knowledge to support all process activities and to facilitate decision making. Plan for the identification, gathering, organising, maintaining, use and retirement of knowledge.				
Process Purpose Statement Provide the knowledge required to support all staff in their work activities and for informed decision making and enhanced productivity.				
Key Management Practices				
BAI08.01 Nurture and facilitate a knowledge-sharing culture. Devise and implement a scheme to nurture and facilitate a knowledge-sharing culture.	**Both**	**Both**	**Both**	BCR-04
BAI08.02 Identify and classify sources of information. Identify, validate and classify diverse sources of internal and external information required to enable effective use and operation of business processes and IT services.	**Both**	**Both**	**Both**	BCR-04
BAI08.03 Organise and contextualise information into knowledge. Organise information based upon classification criteria. Identify and create meaningful relationships between information elements and enable use of information. Identify owners and define and implement levels of access to knowledge resources.	**Both**	**Both**	**Both**	BCR-04
BAI08.04 Use and share knowledge. Propagate available knowledge resources to relevant stakeholders and communicate how these resources can be used to address different needs (e.g.,problem solving, learning, strategic planning and decision making).	**Both**	**Both**	**Both**	BCR-04
BAI08.05 Evaluate and retire information. Measure the use and evaluate the currency and relevance of information. Retire obsolete information.	**Both**	**Both**	**Both**	BCR-04

Figure 35—COBIT 5 Process Practices and the Cloud *(cont.)*

COBIT 5 Process Practice	Main Responsibility			Mapping to CSA CCM V3 Control ID
	SaaS	PaaS	IaaS	
Build, Acquire and Implement *(cont.)*				
BAI09 Manage Assets Manage IT assets through their life cycle to make sure that their use delivers value at optimal cost, they remain operational (fit for purpose), they are accounted for and physically protected, and those assets that are critical to support service capability are reliable and available. Manage software licences to ensure that the optimal number are acquired, retained and deployed in relation to required business usage, and the software installed is in compliance with licence agreements.				
Process Purpose Statement Account for all IT assets and optimise the value provided by these assets.				
Key Management Practices				
BAI09.01 Identify and record current assets. Maintain an up-to-date and accurate record of all IT assets required to deliver services and ensure alignment with configuration management and financial management.	**CSP**	**CSP**	**CSP**	BCR-12 DSI-01, 02, 04 DCS-01 EKM-02 GRM-02
BAI09.02 Manage critical assets. Identify assets that are critical in providing service capability and take steps to maximise their reliability and availability to support business needs.	**CSP** Client	**CSP**	**CSP** Client	BCR-12 DSI-01, 04 DCS-01 EKM-02, 04
BAI09.03 Manage the asset life cycle. Manage assets from procurement to disposal to ensure that assets are utilised as effectively and efficiently as possible and are accounted for and physically protected.	**CSP**	**CSP**	**CSP**	BCR-12 DSI-01, 04, 08 DCS-01 EKM-02, 04 HRS-01
BAI09.04 Optimise asset costs. Regularly review the overall asset base to identify ways to optimise costs and maintain alignment with business needs.	**Both**	**Both**	**Both**	
BAI09.05 Manage licences. Manage software licences so that the optimal number of licences is maintained to support business requirements and the number of licences owned is sufficient to cover the installed software in use.	**CSP**	**CSP**	**Both**	

Figure 35—COBIT 5 Process Practices and the Cloud *(cont.)*

COBIT 5 Process Practice	Main Responsibility			Mapping to CSA CCM V3 Control ID
	SaaS	PaaS	IaaS	
Build, Acquire and Implement *(cont.)*				
BAI10 Manage Assets Define and maintain descriptions and relationships between key resources and capabilities required to deliver IT-enabled services, including collecting configuration information, establishing baselines, verifying and auditing configuration information, and updating the configuration repository.				
Process Purpose Statement Provide sufficient information about service assets to enable the service to be effectively managed, assess the impact of changes and deal with service incidents.				
Key Management Practices				
BAI10.01 Establish and maintain a configuration model. Establish and maintain a logical model of the services, assets and infrastructure and how to record configuration items (CIs) and the relationships amongst them. Include the CIs considered necessary to manage services effectively and to provide a single reliable description of the assets in a service.	**CSP**	**Both**	**Both**	BCR-04, 09 CCC-04 DSI-02 GRM-01 IVS-01, 04 MOS-09
BAI10.02 Establish and maintain a configuration repository and baseline. Establish and maintain a configuration management repository and create controlled configuration baselines.	**CSP**	**Both**	**Both**	BCR-04, 09 CCC-04 DSI-02 GRM-01 IVS-01, 04 MOS-09
BAI10.03 Maintain and control configuration items. Maintain an up-to-date repository of configuration items by populating with changes.	**CSP**	**Both**	**Both**	BCR-04, 09 CCC-04 DSI-02, IVS-01, 02 MOS-09
BAI10.04 Produce status and configuration reports. Define and produce configuration reports on status changes of configuration items.	**CSP**	**Both**	**Both**	BCR-04 CCC-04 DSI-02 IVS-02
BAI10.05 Verify and review integrity of the configuration repository. Periodically review the configuration repository and verify completeness and correctness against the desired target.	**CSP**	**Both**	**Both**	BCR-04 CCC-04 DSI-02

Figure 35—COBIT 5 Process Practices and the Cloud *(cont.)*

COBIT 5 Process Practice	Main Responsibility			Mapping to CSA CCM V3 Control ID
	SaaS	PaaS	IaaS	
Deliver, Service and Support				
DSS01 Manage Operations Co-ordinate and execute the activities and operational procedures required to deliver internal and outsourced IT services, including the execution of pre-defined standard operating procedures and the required monitoring activities.				
Process Purpose Statement Deliver IT operational service outcomes as planned.				
Key Management Practices				
DSS01.01 Perform operational procedures. Maintain and perform operational procedures and operational tasks reliably and consistently.	**Both**	**Both**	**Both**	BCR-04, 11 DSI-05, 08 DCS-02 GRM-03 IVS-03 TVM-02
DSS01.02 Manage outsourced IT services. Manage the operation of outsourced IT services to maintain the protection of enterprise information and reliability of service delivery.	**CSP**	**CSP** Client	**Both**	DCS-05 TVM-02
DSS01.03 Monitor IT infrastructure. Monitor the IT infrastructure and related events. Store sufficient chronological information in operations logs to enable the reconstruction, review and examination of the time sequences of operations and the other activities surrounding or supporting operations.	**CSP**	**CSP** Client	**Both**	BCR-03, 05 IVS-01 SEF-02, 04 TVM-02
DSS01.04 Manage the environment. Maintain measures for protection against environmental factors. Install specialised equipment and devices to monitor and control the environment.	**CSP**	**CSP**	**Both**	BCR-03, 05, 06, 08 DCS-06
DSS01.05 Manage facilities. Manage facilities, including power and communications equipment, in line with laws and regulations, technical and business requirements, vendor specifications, and health and safety guidelines.	**CSP**	**CSP**	**CSP**	BCR-03, 05, 06, 08 DCS-02, 06

Figure 35—COBIT 5 Process Practices and the Cloud *(cont.)*

COBIT 5 Process Practice	Main Responsibility			Mapping to CSA CCM V3 Control ID
	SaaS	PaaS	IaaS	
Deliver, Service and Support *(cont.)*				
DSS02 Manage Service Requests and Incidents Provide timely and effective response to user requests and resolution of all types of incidents. Restore normal service; record and fulfill user requests; and record, investigate, diagnose, escalate and resolve incidents.				
Process Purpose Statement Achieve increased productivity and minimise disruptions through quick resolution of user queries and incidents.				
Key Management Practices				
DSS02.01 Define incident and service request classification schemes. Define incident and service request classification schemes and models.	**CSP**	**CSP** Client	**Both**	IVS-0 SEF-02 through 04
DSS02.02 Record, classify and prioritize requests and incidents. Identify, record and classify service requests and incidents, and assign a priority according to business criticality and service agreements.	**CSP**	**CSP** Client	**Both**	IVS-12 SEF-02, 04
DSS02.03 Verify, approve and fulfill service requests. Select the appropriate request procedures and verify that the service requests fulfill defined request criteria. Obtain approval, if required, and fulfill the requests.	**CSP**	**CSP** Client	**Both**	
DSS02.04 Investigate, diagnose and allocate incidents. Identify and record incident symptoms, determine possible causes and allocate for resolution.	**CSP**	**CSP** Client	**Both**	SEF-02, 04
DSS02.05 Resolve and recover from incidents. Document, apply and test the identified solutions or workarounds and perform recovery actions to restore the IT-related service.	**CSP**	**CSP** Client	**Both**	SEF-02, 04
DSS02.06 Close service requests and incidents. Verify satisfactory incident resolution and/or request fulfillment, and close.	**CSP**	**CSP** Client	**Both**	SEF-02, 04
DSS02.07 Track status and produce reports. Regularly track, analyze and report incident and request fulfillment trends to provide information for continual improvement.	**CSP**	**CSP** Client	**Both**	STA-02

Figure 35—COBIT 5 Process Practices and the Cloud *(cont.)*

COBIT 5 Process Practice	Main Responsibility			Mapping to CSA CCM V3 Control ID
	SaaS	PaaS	IaaS	
Deliver, Service and Support *(cont.)*				
DSS03 Manage Problems Identify and classify problems and their root causes and provide timely resolution to prevent recurring incidents. Provide recommendations for improvements.				
Process Purpose Statement Increase availability, improve service levels, reduce costs, and improve customer convenience and satisfaction by reducing the number of operational problems.				
Key Management Practices				
DSS03.01 Identify and classify problems. Define and implement criteria and procedures to report problems identified, including problem classification, categorisation and prioritisation.	**CSP** Client	**CSP** Client	**Both**	
DSS03.02 Investigate and diagnose problems. Investigate and diagnose problems using relevant subject matter experts to assess and analyse root causes.	**CSP**	**CSP** Client	**Both**	
DSS03.03 Raise known errors. As soon as the root causes of problems are identified, create known-error records and an appropriate workaround, and identify potential solutions.	**CSP**	**CSP** Client	**Both**	
DSS03.04 Resolve and close problems. Identify and initiate sustainable solutions addressing the root cause, raising change requests via the established change management process if required to resolve errors. Ensure that the personnel affected are aware of the actions taken and the plans developed to prevent future incidents from occurring.	**CSP** Client	**CSP** Client	**Both**	
DSS03.05 Perform proactive problem management. Collect and analyse operational data (especially incident and change records) to identify emerging trends that may indicate problems. Log problem records to enable assessment.	**CSP**	**CSP** Client	**Both**	BCR-07 TVM-02

Figure 35—COBIT 5 Process Practices and the Cloud *(cont.)*

COBIT 5 Process Practice	Main Responsibility			Mapping to CSA CCM V3 Control ID
	SaaS	PaaS	IaaS	
Deliver, Service and Support *(cont.)*				
DSS04 Manage Continuity Establish and maintain a plan to enable the business and IT to respond to incidents and disruptions in order to continue operation of critical business processes and required IT services and maintain availability of information at a level acceptable to the enterprise.				
Process Purpose Statement Continue critical business operations and maintain availability of information at a level acceptable to the enterprise in the event of a significant disruption.				
Key Management Practices				
DSS04.01 Define the business continuity policy, objectives and scope. Define business continuity policy and scope aligned with enterprise and stakeholder objectives.	**Both**	**Both**	**Both**	BCR-01, 08, 09, 10, 12 DCS-06
DSS04.02 Maintain a continuity strategy. Evaluate business continuity management options and choose a cost-effective and viable continuity strategy that will ensure enterprise recovery and continuity in the face of a disaster or other major incident or disruption.	**Both**	**Both**	**Both**	BCR-01, 08, 09, 10, 12
DSS04.03 Develop and implement a business continuity response. Develop a business continuity plan (BCP) based on the strategy that documents the procedures and information in readiness for use in an incident to enable the enterprise to continue its critical activities.	**Both**	**Both**	**Both**	BCR-01, 03, 08, 10, 12 DCS-06
DSS04.04 Exercise, test and review the BCP. Test the continuity arrangements on a regular basis to exercise the recovery plans against predetermined outcomes and to allow innovative solutions to be developed and help to verify over time that the plan will work as anticipated.	**Both**	**Both**	**Both**	BCR-02, 12

Figure 35—COBIT 5 Process Practices and the Cloud *(cont.)*

COBIT 5 Process Practice	Main Responsibility			Mapping to CSA CCM V3 Control ID
	SaaS	PaaS	IaaS	
Deliver, Service and Support *(cont.)*				
DSS04.05 Review, maintain and improve the continuity plan. Conduct a management review of the continuity capability at regular intervals to ensure its continued suitability, adequacy and effectiveness. Manage changes to the plan in accordance with the change control process to ensure that the continuity plan is kept up to date and continually reflects actual business requirements.	**Both**	**Both**	**Both**	BCR-01
DSS04.06 Conduct continuity plan training. Provide all concerned internal and external parties with regular training sessions regarding the procedures and their roles and responsibilities in case of disruption.	**Both**	**Both**	**Both**	
DSS04.07 Manage backup arrangements. Maintain availability of business-critical information.	**Both**	**Both**	**Both**	BCR-12 DSI-01, 04 DCS-01 IPY-02 SEF-05
DSS04.08 Conduct post-resumption review. Assess the adequacy of the BCP following the successful resumption of business processes and services after a disruption.	**Both**	**Both**	**Both**	

Figure 35—COBIT 5 Process Practices and the Cloud *(cont.)*

COBIT 5 Process Practice	Main Responsibility			Mapping to CSA CCM V3 Control ID
	SaaS	PaaS	IaaS	
Deliver, Service and Support *(cont.)*				
DSS05 Manage Security Services Protect enterprise information to maintain the level of information security risk acceptable to the enterprise in accordance with the security policy. Establish and maintain information security roles and access privileges and perform security monitoring.				
Process Purpose Statement Minimise the business impact of operational information security vulnerabilities and incidents.				
Key Management Practices				
DSS05.01 Protect against malware. Implement and maintain preventive, detective and corrective measures in place (especially up-to-date security patches and virus control) across the enterprise to protect information systems and technology from malware (e.g., viruses, worms, spyware, spam).	**CSP**	**CSP**	**Both**	AIS-02 DSI-03, 05 HRS-06 IVS-07 MOS-17 TVM-01 through 03
DSS05.02 Manage network and connectivity security. Use security measures and related management procedures to protect information over all methods of connectivity.	**CSP**	**CSP**	**Both**	AIS-02, 04 DCS-03 DSI-03, 05 EKM-03 HRS-06 IAM-03, 04 IVS-06, 08 through 12 TVM-03
DSS05.03 Manage endpoint security. Ensure that endpoints (e.g., laptop, desktop, server, and other mobile and network devices or software) are secured at a level that is equal to or greater than the defined security requirements of the information processed, stored or transmitted.	**Both**	**Both**	**Both**	AIS-02 CCC-04 DCS-03 DSI-03, 05 EKM-03 HRS-06, 12 IAM-01, 03 IVS-05, 07, 12 MOS-11, 12, 14, 16 through 19 TVM-02, 03

Figure 35—COBIT 5 Process Practices and the Cloud *(cont.)*				
COBIT 5 Process Practice	Main Responsibility			Mapping to CSA CCM V3 Control ID
	SaaS	PaaS	IaaS	
Deliver, Service and Support *(cont.)*				
DSS05.04 Manage user identity and logical access. Ensure that all users have information access rights in accordance with their business requirements and co-ordinate with business units that manage their own access rights within business processes.	**Both**	**Both**	**Both**	AIS-02 CCC-04 DSI-01, 03 through 05 DCS-01, 09 EKM-01 HRS-09 IAM-02, 04 through 12 IVS-11, 12 STA-01 TVM-03
DSS05.05 Manage physical access to IT assets. Define and implement procedures to grant, limit and revoke access to premises, buildings and areas according to business needs, including emergencies. Access to premises, buildings and areas should be justified, authorised, logged and monitored. This should apply to all persons entering the premises, including staff, temporary staff, clients, vendors, visitors or any other third party.	**CSP**	**CSP**	**CSP** Client	AIS-02 CCC-04 DSI-01, 03 through 05 DCS-01, 02, 07 through 09 IAM-01 through 03, 13 IVS-08 through 10, 12 MOS-11, 14, 18, 19
DSS05.06 Manage sensitive documents and output devices. Establish appropriate physical safeguards, accounting practices and inventory management over sensitive IT assets, such as special forms, negotiable instruments, special-purpose printers or security tokens.	**Both**	**Both**	**Both**	AIS-02 DSI-03, 05 EKM-01 IAM-02
DSS05.07 Monitor the infrastructure for security-related events. Using intrusion detection tools, monitor the infrastructure for unauthorised access and ensure any events are integrated with general event monitoring and incident management.	**CSP**	**CSP**	**Both**	AAC-02 AIS-02 CCC-04 DSI-03, 05 HRS-06 IAM-07 IVS-01, 12 TVM-02

Figure 35—COBIT 5 Process Practices and the Cloud *(cont.)*

COBIT 5 Process Practice	Main Responsibility			Mapping to CSA CCM V3 Control ID
	SaaS	PaaS	IaaS	
Deliver, Service and Support *(cont.)*				
DSS06 Manage Business Process Controls Define and maintain appropriate business process controls to ensure that information related to and processed by in-house or outsourced business processes satisfies all relevant information control requirements. Identify the relevant information control requirements and manage and operate adequate controls to ensure that information and information processing satisfy these requirements.				
Process Purpose Statement Maintain information integrity and the security of information assets handled within business processes in the enterprise or outsourced.				
Key Management Practices				
DSS06.01 Align control activities embedded in business processes with enterprise objectives. Continually assess and monitor the execution of the business process activities and related controls, based on enterprise risk, to ensure that the processing controls are aligned with business needs.	**Both**	**Both**	**Both**	DSI-03
DSS06.02 Control the processing of information. Operate the execution of the business process activities and related controls, based on enterprise risk, to ensure that information processing is valid, complete, accurate, timely, and secure (i.e., reflects legitimate and authorised business use).	**Both**	**Both**	**Both**	AIS-03 DSI-03, 05
DSS06.03 Manage roles, responsibilities, access privileges and levels of authority. Manage the business roles, responsibilities, levels of authority and segregation of duties needed to support the business process objectives. Authorise access to any information assets related to business information processes, including those under the custody of the business, IT and third parties. This ensures that the business knows where the data are and who is handling data on its behalf.	**Both**	**Both**	**Both**	CCC-04 DCS-02, 08, 09 DSI-03 EKM-01 HRS-06 IAM-02, 05 through 12 IVS-11, 12 STA-01

Figure 35—COBIT 5 Process Practices and the Cloud *(cont.)*

COBIT 5 Process Practice	Main Responsibility			Mapping to CSA CCM V3 Control ID
	SaaS	PaaS	IaaS	
Deliver, Service and Support *(cont.)*				
DSS06.04 Manage errors and exceptions. Manage business process exceptions and errors and facilitate their correction. Include escalation of business process errors and exceptions and the execution of defined corrective actions. This provides assurance of the accuracy and integrity of the business information process.	**CSP**	**CSP**	**Both**	AIS-03 DSI-03, 05
DSS06.05 Ensure traceability of Information events and accountabilities. Ensure that business information can be traced to the originating business event and accountable parties. This enables traceability of the information through its life cycle and related processes. This provides assurance that information that drives the business is reliable and has been processed in accordance with defined objectives.	**CSP**	**CSP**	**Both**	DSI-03 IVS-01
DSS06.06 Secure information assets. Secure information assets accessible by the business through approved methods, including information in electronic form (such as methods that create new assets in any form, portable media devices, user applications and storage devices), information in physical form (such as source documents or output reports) and information during transit. This benefits the business by providing end-to-end safeguarding of information.	**Both**	**Both**	**Both**	AIS-04 DSI-01, 03, 04, 05 DCS-01, 02 EKM-01, 03 HRS-06, 09, 12 IAM-02 through 04, 07 through 12 IVS-05 through 12 MOS-11 STA-01

Figure 35—COBIT 5 Process Practices and the Cloud *(cont.)*

COBIT 5 Process Practice	Main Responsibility			Mapping to CSA CCM V3 Control ID
	SaaS	PaaS	IaaS	
Monitor, Evaluate and Assess				
MEA01 Monitor, Evaluate and Assess Performance and Conformance Collect, validate and evaluate business, IT and process goals and metrics. Monitor that processes are performing against agreed-on performance and conformance goals and metrics and provide reporting that is systematic and timely.				
Process Purpose Statement Provide transparency of performance and conformance and drive achievement of goals.				
Key Management Practices				
MEA01.01 Establish a monitoring approach. Engage with stakeholders to establish and maintain a monitoring approach to define the objectives, scope and method for measuring business solution and service delivery and contribution to enterprise objectives. Integrate this approach with the corporate performance management system.	**CSP**	**CSP** Client	**Both**	STA-04, 06, 08
MEA01.02 Set performance and conformance targets. Work with the stakeholders to define, periodically review, update and approve performance and conformance targets within the performance measurement system.	**CSP**	**CSP** Client	**Both**	STA-04, 06, 08
MEA01.03 Collect and process performance and conformance data. Collect and process timely and accurate data aligned with enterprise approaches.	**CSP**	**CSP** Client	**Both**	IAM-10 through 12 STA-04, 06, 08
MEA01.04 Analyse and report performance. Periodically review and report performance against targets, using a method that provides a succinct all-around view of IT performance and fits within the enterprise monitoring system.	**CSP**	**CSP** Client	**Both**	STA-04, 06, 08
MEA01.05 Ensure the implementation of corrective actions. Assist stakeholders in identifying, initiating and tracking corrective actions in order to address anomalies.	**CSP**	**CSP** Client	**Both**	STA-04, 06, 08

Figure 35—COBIT 5 Process Practices and the Cloud *(cont.)*

COBIT 5 Process Practice	Main Responsibility			Mapping to CSA CCM V3 Control ID
	SaaS	PaaS	IaaS	
Monitor, Evaluate and Assess *(cont.)*				
MEA02 Monitor, Evaluate and Assess the System of Internal Control Continuously monitor and evaluate the control environment, including self-assessments and independent assurance reviews. Enable management to identify control deficiencies and inefficiencies and to initiate improvement actions. Plan, organize and maintain standards for internal control assessment and assurance activities.				
Process Purpose Statement Obtain transparency for key stakeholders on the adequacy of the system of internal controls and thus provide trust in operations, confidence in the achievement of enterprise objectives and an adequate understanding of residual risk.				
Key Management Practices				
MEA02.01 Monitor internal controls. Continuously monitor, benchmark and improve the IT control environment and control framework to meet organisational objectives.	**CSP**	**CSP**	**Both**	AAC-01 GRM-01 STA-04, 08, 09
MEA02.02 Review business process controls effectiveness. Review the operation of controls, including a review of monitoring and test evidence, to ensure that controls within business processes operate effectively. Include activities to maintain evidence of the effective operation of controls through mechanisms such as periodic testing of controls, continuous controls monitoring, independent assessments, command and control centres, and network operations centres. This provides the business with the assurance of control effectiveness to meet requirements related to business, regulatory and social responsibilities.	**CSP**	**CSP** Client	**Both**	AAC-01 STA-04, 08
MEA02.03 Perform control self-assessments. Encourage management and process owners to take positive ownership of control improvement through a continuing programme of self-assessment to evaluate the completeness and effectiveness of management's control over processes, policies and contracts.	**CSP**	**CSP** Client	**Both**	STA-04, 08
MEA02.04 Identify and report control deficiencies. Identify control deficiencies and analyse and identify their underlying root causes. Escalate control deficiencies and report to stakeholders.	**CSP** Client	**CSP** Client	**Both**	STA-04, 08

Figure 35—COBIT 5 Process Practices and the Cloud *(cont.)*

COBIT 5 Process Practice	Main Responsibility			Mapping to CSA CCM V3 Control ID
	SaaS	PaaS	IaaS	
Monitor, Evaluate and Assess *(cont.)*				
MEA02.05 Ensure that assurance providers are independent and qualified. Ensure that the entities performing assurance are independent from the function, groups or organisations in scope. The entities performing assurance should demonstrate an appropriate attitude and appearance, competence in the skills and knowledge necessary to perform assurance, and adherence to codes of ethics and professional standards.	**Client**	**Client**	**Client**	STA-04, 08
MEA02.06 Plan assurance initiatives. Plan assurance initiatives based on enterprise objectives and strategic priorities, inherent risk, resource constraints, and sufficient knowledge of the enterprise.	**CSP**	**CSP** Client	**Both**	AAC-01 STA-04, 08
MEA02.07 Scope assurance initiatives. Define and agree with management on the scope of the assurance initiative, based on the assurance objectives.	**CSP**	**CSP** Client	**Both**	AAC-01 STA-04, 08
MEA02.08 Execute assurance initiatives. Execute the planned assurance initiative. Report on identified findings. Provide positive assurance opinions, where appropriate, and recommendations for improvement relating to identified operational performance, external compliance and internal control system residual risk.	**CSP**	**CSP** Client	**Both**	AAC-01 STA-04, 08

Figure 35—COBIT 5 Process Practices and the Cloud *(cont.)*

COBIT 5 Process Practice	Main Responsibility			Mapping to CSA CCM V3 Control ID
	SaaS	PaaS	IaaS	
Monitor, Evaluate and Assess *(cont.)*				
MEA03 Monitor, Evaluate and Assess Compliance with External Requirements Evaluate that IT processes and IT-supported business processes are compliant with laws, regulations and contractual requirements. Obtain assurance that the requirements have been identified and complied with, and integrate IT compliance with overall enterprise compliance.				
Process Purpose Statement Ensure that the enterprise is compliant with all applicable external requirements.				
Key Management Practices				
MEA03.01 Identify external compliance requirements. On a continuous basis, identify and monitor for changes in local and international laws, regulations and other external requirements that must be complied with from an IT perspective.	Both	Both	Both	AIS-01, 04 AAC-02, 03 BCR-10, 12 GRM-09 SEF-01
MEA03.02 Optimise response to external requirements. Review and adjust policies, principles, standards, procedures and methodologies to ensure that legal, regulatory and contractual requirements are addressed and communicated. Consider industry standards, codes of good practice, and good practice guidance for adoption and adaptation.	Both	Both	Both	AIS-01, 04 GRM-09 SEF-01
MEA03.03 Confirm external compliance. Confirm compliance of policies, principles, standards, procedures and methodologies with legal, regulatory and contractual requirements.	Both	Both	Both	SEF-01
MEA03.04 Obtain assurance of external compliance. Obtain and report assurance of compliance and adherence with policies, principles, standards, procedures and methodologies. Confirm that corrective actions to address compliance gaps are closed in a timely manner.	Both	Both	Both	

Page intentionally left blank

Appendix B. Cloud Computing Assurance Program

Introduction

Overview

ISACA developed the IT Assurance Framework (ITAF) as a comprehensive and good-practice-setting model. ITAF provides standards that are designed to be mandatory and that are the guiding principles under which the IT audit and assurance profession operates. The guidelines provide information and direction for the practice of IT audit and assurance.

Purpose

The assurance program is a tool and template to be used as a road map for the completion of a specific assurance process. ISACA has commissioned assurance programs to be developed for use by IT audit and assurance practitioners. This assurance program is intended to be used by IT audit and assurance professionals with the requisite knowledge of the subject matter under review, as described in ITAF standard 1006 Proficiency.

Control Framework

The assurance programs have been developed in alignment with the ISACA COBIT 5 framework, using generally applicable and accepted good practices. The generic assurance program is presented in COBIT 5 for Assurance and ensures integration of all seven enablers in the assurance approach.

Governance, Risk and Control of IT

Governance, risk and control of IT are critical in the performance of any assurance management process. Governance of the process under review is evaluated as part of the policies and management oversight controls. Risk plays an important role in evaluating what to audit and how management approaches and manages risk. Both issues are evaluated in the assurance program. Enablers are the primary evaluation point in the process. The assurance program identifies the enablers and the steps to determine their design and operating effectiveness.

Responsibilities of IT Audit and Assurance Professionals

IT audit and assurance professionals are expected to customize the "IT Audit and Assurance Program for Cloud Computing" template provided in the appendices section, for the environment in which they are performing the assurance engagement. This document is to be used as a review tool and starting point and may be modified by the IT audit and assurance professional; it is not intended to be a checklist or questionnaire. It is assumed that the IT audit and assurance professional has the necessary subject matter expertise that is required to conduct the work (see following paragraph) and is supervised by a professional with the Certified Information Systems Auditor (CISA) designation and/or necessary subject matter expertise to adequately review the work performed.

Minimum Audit Skills

Cloud computing incorporates many IT processes. Because the focus is on information governance, IT management, network, data, contingency and encryption controls, the audit and assurance professional should have the requisite knowledge of these issues. In addition, proficiency in risk assessment, information security components of IT architecture, risk management, and the threats and vulnerabilities of cloud computing and Internet-based data processing is required. Therefore, it is recommended that the audit and assurance professional who is conducting the assessment has the requisite experience and organizational relationships to effectively execute the assurance processes. Because cloud computing is dependent on web services, the auditor should have at least a basic understanding of Organization for the Advancement of Structured Information Standards (OASIS) Web Services Security (WS-Security or WSS) Standards (*www.oasis-open.org*).

Assurance Program Approach

The assurance program table is a template for a detailed assurance work program, which is based on COBIT 5, for cloud computing in any enterprise.

The assurance work program structures an assurance engagement into three major phases, as depicted in **figure 36**.

Figure 36—Generic COBIT 5-based Assurance Engagement Approach

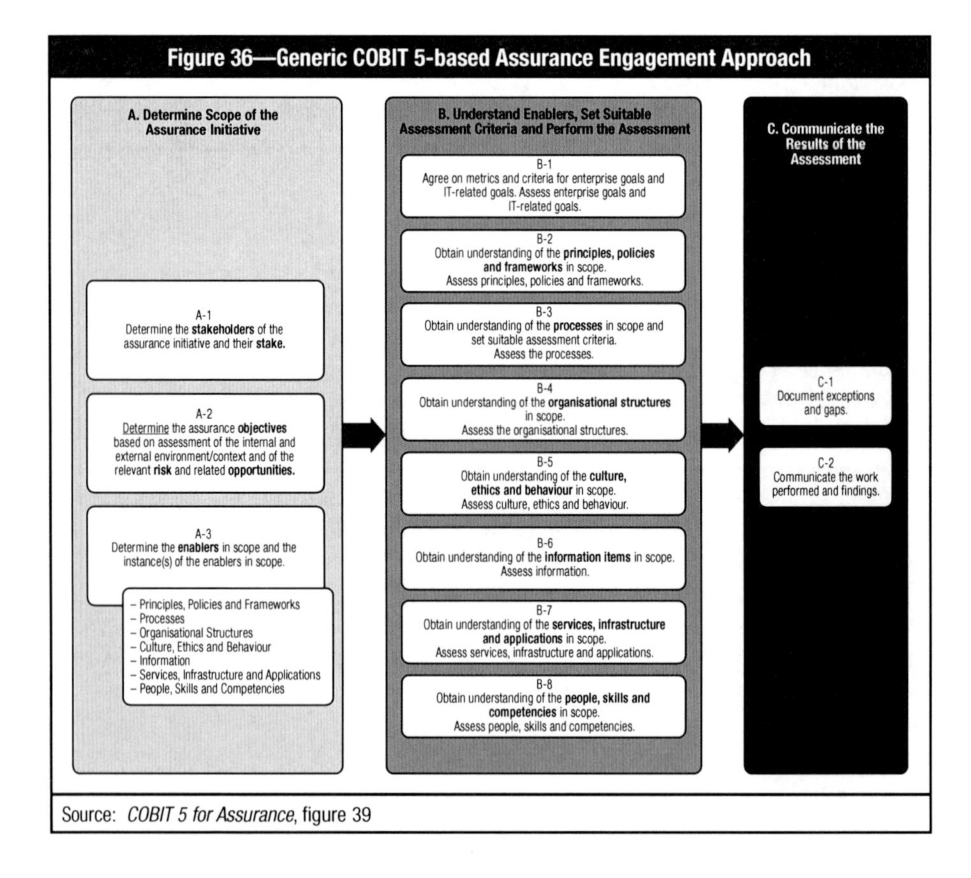

Source: *COBIT 5 for Assurance*, figure 39

As shown in **figure 36**, the proposed assurance engagement approach refers explicitly to all COBIT 5 enabler categories. The COBIT 5 framework explains that the enablers are interconnected, e.g., processes use Organisational Structures and Information items (inputs and outputs). When developing the audit program, it will become clear that when **all** possible entities of **all** enablers are included in the scope and all their detail is reviewed, there is the potential for much duplication.

In the development of the assurance program, care has been taken to avoid such duplications, meaning that:

- Some aspects of a process also relate to another enabler and are classified there, e.g., inputs and outputs are classified under the "Information enabler" heading and treated in detail there.
- Some aspects relating to skills and competencies are to a large extent covered by certain processes, etc.

In practice, audit and assurance professionals, when developing their own customized assurance programs, will also have to make similar decisions to avoid duplication of work.

Generic Assurance Program

The assurance approach depicted in **figure 36** is described in more detail and developed into a **generic assurance program**—including guidance on how to proceed during each step—in *COBIT 5 for Assurance*, Section 2B. The sample assurance program at the end of this appendix is structured fully in line with this generic assurance program. This generic assurance program:

- Is fully aligned with COBIT 5:
  - It explicitly references all seven enablers, i.e., it is no longer exclusively process-focused. It also uses the different dimensions of the enabler model to cover all aspects contributing to the performance of the enablers.
  - It references the COBIT 5 goals cascade to ensure that detailed objectives of the assurance engagement can be put into the enterprise and IT context, and it allows at the same time to link the assurance objectives to enterprise and IT risk and benefits.
- Is quite comprehensive but, at the same time, flexible—the generic program is comprehensive, because it contains assurance steps covering all enablers in much detail. It is flexible, because this detailed structure allows scoping decisions to be made clear and well understood, i.e., the audit and assurance professional can decide to not cover a set of enablers or some enabler instances. Such a decision will reduce the scope and related assurance engagement effort, but it will also be transparent and visible to the assurance engagement user on what is covered and what is not covered.
- Is structured as follows:
  - The table follows the flow described in the **figure 36**, but splits each phase into different steps and substeps.
  - For each step, a short description is included, and guidance for the audit and assurance professional on how to proceed with the step is included in the right column (text in italic).

Customization of the Assurance Program

Customization and completion of the example assurance program in the table will still be required and will consist of refining the scope by selecting goals and enabler instances—the lists included in the example are comprehensive yet still remain an example (i.e., different strategic priorities of the enterprise may dictate a different scope). This list can be considered prohibitive by some, because it can lead to a very broad scope, hence, a very expensive assurance engagement, meaning that a selection and prioritization will be required. The audit and assurance professional must consider the following steps:

- Determine the stakeholders of the assurance initiative and their stakes.
- Determine the assurance objectives based on assessment of the internal and external environment/context, including the strategic objectives and priorities of the enterprise.
- Determine the **enablers** in scope and the instance(s) of the enablers in scope.

In each phase, one or two enabler examples are fully elaborated, to illustrate and demonstrate the suggested approach. The assurance program phases for the other processes and other enablers in scope need to be detailed to the required level of detail.

Using the Assurance Program

In the following section, the cloud computing assurance topic is fully developed based on the generic assurance program. This detailed program contains the following additional information:

- In the "guidance" column, the shaded text is specific for the example and provides practical guidance, e.g., examples on which processes to include in scope, on which organizational structures to include in scope, on how to set assessment criteria for the different enablers, on how to actually assess the different enablers.
- Two additional columns, allowing the audit and assurance professional to identify and cross-reference issues and to record comments.
  - **Cross reference issues**—This column can be used to flag a finding/issue that the IT audit and assurance professional wants to further investigate or establish as a potential finding. The potential findings should be documented in a work paper that indicates the disposition of the findings (formally reported, reported as a memo or verbal finding, or waived).
  - **Record comments**

The example provides a partially elaborated assurance program for an assurance engagement of cloud computing arrangements in an enterprise. For most of the enablers, there are several instances in scope, e.g., in this example, six key processes and five organizational structures are identified to be in scope. However, the assurance program—by means of examples—has been fully developed for only one instance of each enabler. The remaining instances can be deduced very similarly to the examples, using the COBIT 5 framework and the *COBIT 5: Enabling Processes* guides.

Assurance Engagement: Cloud Computing

Assurance Topic

The topic covered by this assurance engagement is cloud computing (from a user perspective).

Goal of the Review

The goal of the review is to provide assurance for all audit stakeholders on the appropriate design and operational effectiveness of the COBIT 5 enablers, relevant to cloud computing. The enablers that are specifically relevant to cloud computing depend greatly on the additional business impact and risk that using cloud services (compared to traditional outsourcing) represents. This additional cloud computing risk has following main components:

- Greater dependency on third parties:
  - Increased vulnerabilities in external interfaces
  - Increased risk in aggregated data centers
  - Immaturity of the service providers with the potential for service provider ongoing concern issues
  - Increased reliance on independent assurance processes
- Increased complexity of compliance with laws and regulations:
  - Greater magnitude of privacy risk
  - Transborder flow of personally identifiable information (PII)
  - Affecting contractual compliance
- Reliance on the Internet as the primary conduit to the enterprise's data introduces:
  - Security issues with a public environment
  - Availability issues of Internet connectivity
- Due to the dynamic nature of cloud computing:
  - The location of the processing facility may change according to load balancing
  - The processing facility may be located across international boundaries
  - Operating facilities may be shared with competitors
  - Legal issues (liability, ownership, etc.) relating to differing laws in hosting countries may put data at risk

As a summary, **this assurance program focus is on**:

- The governance affecting cloud computing
- The contractual compliance between the service provider and customer
- Privacy and regulation issues concerning cloud computing
- Other specific cloud computing attention points

This cloud computing assurance review is not designed to provide assurance on the design and operational effectiveness of the cloud computing service provider's internal controls, as this basically requires covering all COBIT 5 enablers and thus conducting a full scope assurance program. The assurance review is neither designed to replace or focus on audits that provide assurance of specific application processes and excludes assurance of an application's functionality and suitability.

Scoping

The scope of the assurance engagement is expressed in the functions of the seven COBIT 5 enablers, according to the following audit program table.

Some enabler instances are "standard" COBIT 5, i.e., they are described in varying degrees of detail in the COBIT 5 framework or *COBIT 5: Enabling Processes*. This includes mainly COBIT 5 processes, but also the enabler examples included in this or similar publications.

Other enabler instances are more specific and are not described in COBIT 5, yet they are valid instances of an enabler category. The meaning of this distinction is that for the COBIT 5 items, reference material is freely available and customizable; whereas, for the "nonstandard" COBIT 5 items, other references may need to be used and/or more preparation or customization work will be required.

Because the areas under review rely heavily on the effectiveness of core IT general controls, it is recommended that audit/assurance reviews of the following areas be performed prior to the execution of the cloud computing review, so that appropriate reliance can be placed on these assessments:

- Identity management (if the enterprise's identity management system is integrated with the cloud computing system)
- Security incident management (to interface with and manage cloud computing incidents)
- Network perimeter security (as an access point to the Internet)
- Systems development (in which the cloud is part of the application infrastructure)
- Project management
- IT risk management
- Data management (for data transmitted and stored on cloud systems)
- Vulnerability management

Figure 37—IT Audit and Assurance Program for Cloud Computing

A—Determine Scope of the Assurance Initiative.

Ref.	Assurance Step	Guidance		Issue Cross-reference	Comment
A-1	Determine the **stakeholders** of the assurance initiative and their **stakes**.				
A-1.1	Identify the intended users of the assurance report and their stakes in the assurance engagement. This is the assurance objective.	**Intended user(s) of the assurance report**	**Board/audit committee:** Needs assurance over the effectiveness and efficiency of cloud computing processes within the enterprise, and, in this particular example, over the governance affecting cloud computing, the contractual compliance between the service provider and the enterprise, and any regulations and security/privacy issues.		
A-1.2	Identify the interested party accountable and responsible for the subject matter over which assurance needs to be provided.	**Accountable and responsible parties for the subject matter**	**Steering committee:** Accountable for guidance of the cloud computing services, including management and monitoring of the services, allocation of resources, delivery of benefits and value, and management of risk. **Business executives:** The individuals responsible for identifying requirements, approving design and managing performance. These people are, together with IT management, responsible for managing the correct and controlled use of the cloud computing services—in line with good practices. **IT management:** Responsible for managing the correct and controlled use of cloud computing services—together with the business executives.		

Figure 37—IT Audit and Assurance Program for Cloud Computing *(cont.)*

Ref.	Assurance Step	Guidance	Issue Cross-reference	Comment
A—Determine Scope of the Assurance Initiative. *(cont.)*				
A-2	Determine the assurance **objectives** based on assessment of the internal and external environment/context and of the relevant **risk** and related **opportunities** (i.e., not achieving the enterprise goals).	Assurance objectives are essentially a more detailed and tangible expression of those enterprise objectives relevant to the subject of the assurance engagement. Enterprise objectives can be formulated in terms of the generic enterprise goals (COBIT 5 framework) or they can be expressed more specifically. **Objectives of the assurance engagement can be expressed using the COBIT 5 enterprise goals, the IT-related goals (which relate more to technology), information goals or any other set of specific goals.** **Objectives of the assurance engagement will consider all three value objective components, i.e., delivering benefits that support strategic objectives, optimizing the risk that strategic objectives are not achieved and optimizing resource levels required to achieve the strategic objectives.**		
A-2.1	Understand the enterprise strategy and priorities.	*Inquire with executive management or through available documentation (corporate strategy, annual report, etc.) about the enterprise strategy and priorities for the coming period, and document them to the extent the process under review is relevant.*		
A-2.2	Understand the internal context of the enterprise.	*Identify all internal environmental factors that could influence the **performance of the process under review.***		
A-2.3	Understand the external context of the enterprise.	*Identify all external environmental factors that could influence the **performance of the process under review.***		

Figure 37—IT Audit and Assurance Program for Cloud Computing *(cont.)*

Ref.	Assurance Step	Guidance		Issue Cross-reference	Comment
A—Determine Scope of the Assurance Initiative. *(cont.)*					
A-2.4	Given the overall assurance objective, translate the identified strategic priorities into concrete objectives for the assurance engagement.	The following goals are retained as key goals to be supported, in reflection of enterprise strategy and priorities:			
		Key goals	Enterprise goals: • EG03 Managed business risk (safeguarding of assets) • EG04 Compliance with externals laws and regulations IT-related goals: • ITG02 IT compliance and support for business compliance with external laws and regulations • ITG04 Managed IT-related business risk • ITG10 Security of information, processing infrastructure and applications		

Figure 37—IT Audit and Assurance Program for Cloud Computing *(cont.)*

A—Determine Scope of the Assurance Initiative. *(cont.)*					
Ref.	**Assurance Step**	**Guidance**		**Issue Cross-reference**	**Comment**
A-2.4 *(cont.)*		**Additional goals**	Enterprise goals: • EG01 Stakeholder value of business investments • EG08 Agile responses to a changing business environment • EG10 Optimisation of service delivery costs IT-related goals: • ITG05 Realised benefits from IT-enabled investments and services portfolio • ITG07 Delivery of IT services in line with business requirements • ITG09 IT agility • ITG11 Optimisation of IT assets, resources and capabilities		
A-2.5	Define the organizational boundaries of the assurance initiative.	*Describe the organizational boundaries of the assurance engagement, i.e., to which organizational entities the review is limited. All other aspects of scope limitation are identified during phase A-3.*			

Figure 37—IT Audit and Assurance Program for Cloud Computing *(cont.)*

A—Determine Scope of the Assurance Initiative. *(cont.)*				
Ref.	**Assurance Step**	**Guidance**	**Issue Cross-reference**	**Comment**
A-3	Determine the **enablers** in scope and the instance(s) of the enablers in scope.	COBIT 5 identifies seven enabler categories. In this section all seven are covered, and the assurance professional has the opportunity to select enablers from all categories to obtain the most comprehensive scope for the assurance engagement.		
A-3.1	Define the **Principles, Policies and Frameworks** in scope.	Guiding principles and policies include: • ISMS policy (output, APO13.01) • Connectivity security policy (output, DSS05.02; input, APO01.04) • Information architecture model (input, DSS05.04; output, APO03.02) • Legal and regulatory compliance requirements (input, MEA03.01)		

Figure 37—IT Audit and Assurance Program for Cloud Computing *(cont.)*

A—Determine Scope of the Assurance Initiative. *(cont.)*					
Ref.	**Assurance Step**	**Guidance**		**Issue Cross-reference**	**Comment**
A-3.2	Define which **Processes** are in scope of the review. Processes will be assessed during phase B of the assurance engagement against the criteria defined in phase A, and assessments will typically focus on: • Achievement of process goals • Application of process good practices • Existence and quality of work products (inputs and outputs in so far not covered by the information items assessments)	*COBIT 5: Enabling Processes* distinguishes a "governance" domain with a set of processes and a "management" domain, with four sets of processes. The processes in scope are identified using the goals cascade and subsequent customization. The resulting lists contain key processes and additional processes to be considered during this assurance engagement. Available resources will determine whether they can all be effectively assessed.			
		Key processes	• AP001 Manage the IT management framework. • AP009 Manage service agreements. • AP010 Manage suppliers. • AP012 Manage risk. • AP013 Manage security. • DSS02 Manage service requests and incidents. • DSS05 Manage security services. • MEA03 Monitor, evaluate and assess compliance with external requirements.		
		Additional processes	• EDM03 Ensure risk optimisation. • BAI02 Manage requirements definition. • BAI03 Manage solutions identification and build. • BAI06 Manage changes. • BAI10 Manage configuration. • DSS04 Manage continuity. • MEA01 Monitor, evaluate and assess performance and conformance. • MEA02 Monitor, evaluate and assess the system of internal control.		

Figure 37—IT Audit and Assurance Program for Cloud Computing *(cont.)*

A—Determine Scope of the Assurance Initiative. *(cont.)*

Ref.	Assurance Step	Guidance		Issue Cross-reference	Comment
A-3.3	Define which **Organisational Structures** will be in scope. Organisational Structures will be assessed during phase C of the assurance engagement against the criteria defined in phase B, and assessments will typically focus on: • Achievement of Organisational Structure goals, i.e., decisions • Application of Organisational Structures good practices	Based on the key processes identified in A-3.2, the following Organisational Structures and functions are considered to be in scope of this assurance engagement, and available resources will determine which ones will be reviewed in detail.			
		Key Organisational Structures	• Business executives • Service manager • Chief information officer • Business process owners • Chief information security officer		
		Additional Organisational Structures	• Chief executive officer • Head IT operations • Risk function • Privacy officer • Compliance • Audit		
A-3.4	Define the **Culture, Ethics and Behaviour** aspects in scope.	In the context of this engagement, the following enterprisewide culture and behaviours are in scope: • CSP collaboration is a key success factor. • Management proactively monitors risk and action plan progress. • Policies reflect risk appetite and risk tolerance. • The value of risk is recognized. • Data confidentiality is recognized enterprisewide as an attention point when using cloud computing services.			

Figure 37—IT Audit and Assurance Program for Cloud Computing *(cont.)*

A—Determine Scope of the Assurance Initiative. *(cont.)*					
Ref.	**Assurance Step**	**Guidance**		**Issue Cross-reference**	**Comment**
A-3.5	Define the **Information items** in scope. Information items will be assessed during phase C of the assurance engagement against the criteria defined in phase B, and assessments will typically focus on: • Achievement of Information goals, i.e,. quality criteria of the information items • Application of Information good practices (Information attributes)	*COBIT 5: Enabling Processes* defines a number of inputs and outputs between processes. Based on the fact that APO09, APO10, APO12, APO13, DSS05 and MEA03 were defined as key processes in scope, the related inputs and outputs are considered in this section. Key priorities and availability of resources will determine how many and which ones will be reviewed in detail. The following items are considered for this example.			
		Key Information Items	• IT-related policies (output, APO01.03; input, all APO, all BAI, all DSS, all MEA) • Communication on IT objectives (output APO01.04; input, all APO, all BAI, all DSS, all MEA) • Data classification guidelines (output, APO01.06; input APO03.02, BAI02.01, DSS05.02, DSS06.01) • Data security and control guidelines (output, APO01.06; input, BAI02.01) • Data integrity procedures (output, APO01.06; input, BAI02.01, DSS06.01) • Non-compliance remedial actions (output, APO01.08; input, MEA01.05) • SLAs (output, APO09-03) • Service level performance reports (output, APO09.04) • Evaluations against SLAs (input, APO09.05; output, BAI04.01) • Supplier contracts (input, APO10.01) • Identified supplier delivery risk (output, APO10.04; input, APO12.01, APO12.03, BAI01.01) • Identified contract requirements to minimise risk (output, APO10.04)		

Figure 37—IT Audit and Assurance Program for Cloud Computing *(cont.)*

Ref.	Assurance Step	Guidance		Issue Cross-reference	Comment
A—Determine Scope of the Assurance Initiative. *(cont.)*					
A-3.5 *(cont.)*		**Key Information Items *(cont.)***	• Risk analysis results (output APO12.02; input EDM03.03, APO01.03, APO02.02, BAI01.10) • Review results of third-party risk assessments (output, APO12.04; input, EDM03.03, APO10.04, MEA02.01) • Results of penetration tests (output, DSS05.02; input, MEA02.08) • Identified compliance gaps (output, MEA03.03; input, MEA02.08) • Reports of noncompliance issues and root causes (output, MEA03.04; input, EDM01.03, MEA02.07)		
A-3.5 *(cont.)*		**Additional Information Items**	• Improvement action plans and remediations (output, APO09.04; input, APO02.02, APO08.02) • Satisfaction analyses (input, APO09.04) • Supplier significance and evaluation criteria (output, APO10.01) • Decision results of supplier evaluations (output, APO10.02; input, EDM04.01, BAI02.02) • Supplier compliance monitoring review results (output, APO10.05; MEA01.03) • Supplier roles and responsibilities (output, APO10.03) • IT risk scenarios (output, APO12.02) • Emerging risk issues and factors (output, APO12.01; input, EDM03.01, APO01.03, APO02.02) • Security event logs (output, DSS05.07) • Compliance requirements register (output, MEA03.01)		

Figure 37—IT Audit and Assurance Program for Cloud Computing *(cont.)*

A—Determine Scope of the Assurance Initiative. *(cont.)*				
Ref.	**Assurance Step**	**Guidance**	**Issue Cross-reference**	**Comment**
A-3.6	Define the **Services, Infrastructure and Applications** in scope.	In the context of this assignment, and taking into account the goals identified in A-2.4, following services and related applications or infrastructure could be considered in scope of the review: • Risk management tool • Service provider evaluation tool • SLA monitoring tool		
A-3.7	Define the **People, Skills and Competencies** in scope. Skill sets and competencies will be assessed during phase C of the assurance engagement against the criteria defined in phase B, and assessments will typically focus on: • Achievement of skills set goals • Application of skills set and competencies good practices	In the context of this engagement, taking into account key processes and key roles, the following skill sets are included in scope: • Cloud vendor management skills • Understanding of cloud security and compliance components • Data integration skills		

Figure 37—IT Audit and Assurance Program for Cloud Computing *(cont.)*

B—Understand Enablers, Set Suitable Assessment Criteria and Perform the Assessment			
Ref.	**Assurance Steps and Guidance**	**Issue Cross-reference**	**Comment**
B-1	**Agree on metrics and criteria for enterprise goals and IT-related goals and assess enterprise goals and IT-related goals.**		
B-1.1	Obtain (and agree on) metrics for enterprise goals and expected values of the metrics and assess whether enterprise goals in scope are achieved. *Leverage the list of suggested metrics for the enterprise goals to define, discuss and agree on a set of relevant, customized metrics for the enterprise goals, taking care that the suggested metrics are driven by the performance of the topic of this assurance initiative.* *Next, agree on the expected values for these metrics, i.e., the values against which the assessment will take place.*		
	The following metrics and expected values are agreed on for the key enterprise goals defined in step A-2.4;.		

Enterprise Goal	Metric	Expected Outcome (Ex)	Assessment Step
EG03 Managed business risk (safeguarding of assets)	• Percent of critical business objectives and services covered by risk assessment • Frequency of update of risk profile	A. At least 85 percent of the cloud services are covered by an adequate information security risk assessment B. Have at least a yearly updated information security risk profile on the cloud services used by the enterprise.	Criteria A: (example) • Gather all the information security risk assessments for the cloud services used by the enterprise. • Determine the ratio of cloud services covered by the information risk assessment vs. the total cloud services used by the enterprise.

Figure 37—IT Audit and Assurance Program for Cloud Computing *(cont.)*

Ref.	Assurance Steps and Guidance				Issue Cross-reference	Comment
	B—Understand Enablers, Set Suitable Assessment Criteria and Perform the Assessment *(cont.)*					
B-1.1 *(cont.)*	**Enterprise Goal**	**Metric**	**Expected Outcome (Ex)**	**Assessment Step**		
	EG03 Managed business risk (safeguarding of assets) *(cont.)*			Criteria B: (example) • Obtain the information security risk profile for cloud services used. • Verify the date last updated was less than one year ago.		
	EG04 Compliance with externals laws and regulations	Number of regulatory noncompliance issues relating to contractual agreements with business partners	No contractual noncompliance issues can exist caused by using cloud services instead performing it in-house.	Criteria: (example) • Gather all privacy and security-related contractual obligations from clients. • Verify that the cloud services being used are not violating the terms of the privacy and security-related contractual obligations.		

<table>
<tr><th colspan="7">Figure 37—IT Audit and Assurance Program for Cloud Computing (cont.)</th></tr>
<tr><th colspan="7">B—Understand Enablers, Set Suitable Assessment Criteria and Perform the Assessment (cont.)</th></tr>
<tr><th>Ref.</th><th colspan="4">Assurance Steps and Guidance</th><th>Issue Cross-reference</th><th>Comment</th></tr>
<tr><td rowspan="4">B-1.2</td><td colspan="4"><u>Obtain</u> (and <u>agree</u> on) metrics for IT-related goals and expected values of the metrics and <u>assess</u> whether IT-related goals in scope are achieved.</td><td></td><td></td></tr>
<tr><td colspan="4">The following metrics and expected values are agreed for the key IT-related goals defined in step A-2.4.</td><td></td><td></td></tr>
<tr><th>IT-related Goal</th><th>Metric</th><th>Expected Outcome (Ex)</th><th>Assessment Step</th><td></td><td></td></tr>
<tr><td>ITG02 IT compliance and support for business compliance with external laws and regulations</td><td>Number of cloud services-related noncompliance issues reported to the board or causing public comment or embarrassment</td><td>Maximum of one cloud service related noncompliance issue reported to the board or causing public comment or embarrassment per year</td><td>Criteria: Maximum of one cloud service-related noncompliance issue reported to the board or causing public comment or embarrassment per year (example):
• Obtain an overview of all cloud service-related noncompliance issues in the past year.
• Verify impact analysis per issues.
• Mark issues reported to the board or causing public comment or embarrassment.</td><td></td><td></td></tr>
</table>

Figure 37—IT Audit and Assurance Program for Cloud Computing *(cont.)*

B—Understand Enablers, Set Suitable Assessment Criteria and Perform the Assessment *(cont.)*						
Ref.	**Assurance Steps and Guidance**				**Issue Cross-reference**	**Comment**
B-1.2 *(cont.)*	**IT-related Goal**	**Metric**	**Expected Outcome (Ex)**	**Assessment Step**		
	ITG04 Managed IT-related business risk	• Percent of critical business processes, IT services and IT-enabled business programmes covered by risk assessment • Number of significant IT-related incidents that were not identified in risk assessment • Percent of enterprise risk assessments including IT-related risk • Frequency of update of risk profile	*Agree on the expected values for the IT-related goal metrics, i.e., the values against which the assessment will take place.*	*In this step, the related metrics for each goal will be reviewed and an assessment will be made whether the defined criteria are achieved.*		

Figure 37—IT Audit and Assurance Program for Cloud Computing *(cont.)*

B—Understand Enablers, Set Suitable Assessment Criteria and Perform the Assessment *(cont.)*						
Ref.	**Assurance Steps and Guidance**				**Issue Cross-reference**	**Comment**
B-1.2 *(cont.)*	**IT-related Goal**	**Metric**	**Expected Outcome (Ex)**	**Assessment Step**		
	ITG10 Security of information, processing infrastructure and applications	• Number of security incidents related to cloud services causing financial loss, business disruption or public embarrassment • Time to grant, change and remove access privileges, compared to agreed-on service levels for cloud services • Frequency of cloud services security assessment against latest standards and guidelines	• Maximum of one security incident related to cloud services causing financial loss, business disruption or public embarrassment per year. • Time to grant, change and remove access privileges is never more than one working day for cloud services. • A security assessment of cloud services against latest standards and guidelines is done twice a year.	Criteria: Maximum of one security incident related to cloud services causing financial loss, business disruption or public embarrassment per year (example): • Obtain an overview of all security incidents in the past year related to cloud services. • Verify impact analysis per incident. • Mark incidents that caused financial loss, business disruption or public embarrassment.		

Figure 37—IT Audit and Assurance Program for Cloud Computing *(cont.)*

<table>
<tr><th colspan="5">B—Understand Enablers, Set Suitable Assessment Criteria and Perform the Assessment (cont.)</th></tr>
<tr><th>Ref.</th><th colspan="2">Assurance Steps and Guidance</th><th>Issue Cross-reference</th><th>Comment</th></tr>
<tr><td>B-2</td><td colspan="2">Obtain an understanding of the Principles, Policies and Frameworks in scope and set suitable assessment criteria.
Assess Principles, Policies and Frameworks connectivity security policy.</td><td></td><td></td></tr>
<tr><td colspan="3"><u>Repeat</u> Steps B-2.1 through B-2.5 for all remaining Principles, Policies and Frameworks in scope:
a. Connectivity security policy
b. ISMS policy
c. Information architecture model
d. Legal and regulatory compliance requirements</td><td></td><td></td></tr>
<tr><td>B-2.1a</td><td colspan="2"><u>Understand</u> the Principles, Policies and Frameworks context.</td><td></td><td></td></tr>
<tr><td>B-2.2a</td><td colspan="2"><u>Understand</u> the stakeholders of the Principles, Policies and Frameworks.</td><td></td><td></td></tr>
<tr><td rowspan="2">B-2.3a</td><td colspan="2"><u>Understand</u> the goals for the Principles, Policies and Frameworks, and the related metrics and agree on expected values. Assess whether the Principles, Policies and Frameworks goals (outcomes) are achieved, i.e., assess the effectiveness of the Principles, Policies and Frameworks.</td><td></td><td></td></tr>
<tr><td>In the context of this assurance engagement, one of the relevant policies includes the connectivity security policy. The following goals will be reviewed.</td><td>For the Principles, Policies and Frameworks at hand—connectivity security policy—the assurance professional will perform the following assurance steps.</td><td></td><td></td></tr>
</table>

Figure 37—IT Audit and Assurance Program for Cloud Computing *(cont.)*

B—Understand Enablers, Set Suitable Assessment Criteria and Perform the Assessment *(cont.)*						
Ref.	**Assurance Steps and Guidance**				**Issue Cross-reference**	**Comment**
B-2.3a *(cont.)*	**Process Goal**	**Related Metric**	**Criteria/Expected Value**	**Assessment Step**		
	Comprehensiveness		The policy should be comprehensive in its coverage.	Verify that the policy is comprehensive in its coverage.		
	Currency		The policy is up to date. This would at least require: • A yearly validation of all policies whether they are still up to date • An indication on the policies regarding their expiration date or date of last update	Verify that the agreement is up to date. This would at least require: • A check of when the last update of the policy was done • Verification whether the policy indicates an expiration date and date of last update		
	Flexibility		The policy should be flexible. It should be structured in such a way that it is easy to add or update as circumstances would require it.	Verify the flexibility of the policy, i.e., that it is structured in such a way that it is easy to add or update policies as circumstances would require it.		
	Availability		• The policy is available to all stakeholders. • The policy is easy to navigate and has a logical structure.	• Verify that policy is available to all stakeholders. • Verify that policy is easy to navigate and has a logical and hierarchical structure.		

<table>
<tr><th colspan="6">Figure 37—IT Audit and Assurance Program for Cloud Computing (cont.)</th></tr>
<tr><th colspan="6">B—Understand Enablers, Set Suitable Assessment Criteria and Perform the Assessment (cont.)</th></tr>
<tr><th>Ref.</th><th colspan="3">Assurance Steps and Guidance</th><th>Issue Cross-reference</th><th>Comment</th></tr>
<tr><td rowspan="2">B-2.4a</td><td colspan="3"><u>Understand</u> the life cycle stages of the principles, policies and frameworks, and agree on the relevant criteria.
Assess to what extent the Principles, Policies and Frameworks life cycle is managed</td><td></td><td></td></tr>
<tr><td colspan="2">The life cycle of the IT-related policies is managed by the process APO01. The review of this life cycle is therefore equivalent to a process review of process APO01 Manage the IT management framework.</td><td>In the context of this assurance engagement, process APO01 is not yet in scope, as it was marked as an “additional process.” A review of this process means adding it to the key processes in section A-3.1.</td><td></td><td></td></tr>
<tr><td rowspan="6">B-2.5a</td><td colspan="3"><u>Understand</u> good practice related to the Principles, Policies and Frameworks and expected values.
Assess the Principles, Policies and Frameworks design, i.e., assess to what extent expected good practices are applied.</td><td></td><td></td></tr>
<tr><td colspan="2">In the context of this assurance engagement and the policy at hand, it is expected that the following good practices from the list above are all applied.</td><td>For the Principles, Policies and Frameworks at hand—connectivity security policy—the assurance professional will perform the following assurance steps.</td><td></td><td></td></tr>
<tr><th>Good Practice</th><th>Criteria</th><th>Assessment Step</th><td></td><td></td></tr>
<tr><td>Scope and validity</td><td>The scope is described and the validity date is indicated.</td><td>Verify that the scope of the policy is described and the validity date is indicated.</td><td></td><td></td></tr>
<tr><td>Exception and escalation</td><td>• The exception and escalation procedure is explained and commonly known.
• The exception and escalation procedure has not become the de facto standard procedure.</td><td>• Verify that the exception and escalation procedure for the cloud security connectivity is described, explained and commonly known.
• Through observation of a representative sample, verify that the exception and escalation procedure for the cloud security connectivity has not become the de facto standard procedure.</td><td></td><td></td></tr>
<tr><td>Compliance</td><td>The compliance checking mechanism and non-compliance consequences are clearly described and enforced.</td><td>Verify that the compliance checking mechanism and non-compliance consequences are clearly described and enforced for the cloud security connectivity measures.</td><td></td><td></td></tr>
</table>

Figure 37—IT Audit and Assurance Program for Cloud Computing *(cont.)*

Ref.	Assurance Steps and Guidance	Issue Cross-reference	Comment
B—Understand Enablers, Set Suitable Assessment Criteria and Perform the Assessment *(cont.)*			
B-3	**Obtain understanding of the Processes in scope and set suitable assessment criteria: for each process in scope (as determined in step A-3.1), additional information is collected and assessment criteria are defined. Assess the Processes.**		
Repeat Steps B-3.1 through B-3.7 for **all remaining processes in scope.** a. APO10 Manage Suppliers b. APO01 Manage the IT management framework. c. APO09 Manage service agreements. d. APO12 Manage risk. e. APO13 Manage security. f. DSS02 Manage service requests and incidents. g. DSS05 Manage security services. h. MEA03 Monitor, evaluate and assess compliance with external requirements.			
B-3.1a	Understand the **Process context.**		
	Example: Process APO10 *Manage suppliers* is the key process in the enterprise to select suppliers and manage the supplier relationship and risk. Process description: Manage IT-related services provided by all types of suppliers to meet enterprise requirements, including the selection of suppliers, management of relationships, management of contracts, and reviewing and monitoring of supplier performance for effectiveness and compliance.		
B-3.2a	Understand the **Process purpose.**		
	The purpose of process APO10 is as per the standard COBIT 5 process, i.e., "Minimise the risk associated with non-performing suppliers and ensure competitive pricing."		

<table>
<tr><th colspan="5">Figure 37—IT Audit and Assurance Program for Cloud Computing (cont.)</th></tr>
<tr><th colspan="5">B—Understand Enablers, Set Suitable Assessment Criteria and Perform the Assessment (cont.)</th></tr>
<tr><th>Ref.</th><th colspan="2">Assurance Steps and Guidance</th><th>Issue
Cross-reference</th><th>Comment</th></tr>
<tr><td rowspan="2">B-3.3a</td><td colspan="2">Understand all process stakeholders and their roles, in so far not covered yet by A-3.2.
This is equivalent to understanding the real RACI chart of the process.</td><td></td><td></td></tr>
<tr><td colspan="2">The stakeholders of the process are already defined in the RACI chart as a result of step A-3.2. In addition to those stakeholders, this process relies also on the following function(s), which therefore will need to be involved during the assurance engagement: none.</td><td></td><td></td></tr>
<tr><td rowspan="2">B-3.4a</td><td colspan="2">Understand the Process goals and related metrics and define expected Process values (criteria), and assess whether the Process goals are achieved, i.e., assess the effectiveness of the process.</td><td></td><td></td></tr>
<tr><td>The Process APO10 Manage suppliers has three standard defined process goals, as described in COBIT 5: Enabling Processes, chapter 5, p. 97. Based on these goals and their related metrics, the following concrete goals and associated metrics are defined for this process.</td><td>The following activities can be performed to assess whether the goals are achieved.</td><td></td><td></td></tr>
</table>

Figure 37—IT Audit and Assurance Program for Cloud Computing *(cont.)*

Ref.	Assurance Steps and Guidance				Issue Cross-reference	Comment
B—Understand Enablers, Set Suitable Assessment Criteria and Perform the Assessment *(cont.)*						
B-3.4a *(cont.)*	**Process Goal**	**Related Metrics**	**Criteria/Expected Values**	**Assessment Step**		
	Suppliers perform as agreed.	Percent of CSPs meeting agreed-on requirements	• 80 percent of all CSPs should be meeting all the agreed-on requirements in the SLAs. • Not more than 10 percent of all CSPs can slightly fail to meet an SLA requirement twice a month. • Not more than 10 percent of all CSPs can significantly fail to meet an SLA requirement once a month.	Criteria: • Gather all the service level performance reports and evaluations against SLAs. • Verify the expected values. • For suppliers failing to meet the agreed-on requirements, verify if this is a recurring trend for those suppliers. • Determine the root cause of a trend and suggest actions for improvement of the SLA or suggest considering another supplier.		
	Supplier risk is assessed and properly addressed.	Percent of cloud services risk-related incidents resolved acceptably (time and cost)	99 percent of cloud services risk-related incidents should be resolved in the agreed-on time and cost frame.	Criteria: • Obtain an overview of cloud services-related incidents. • Verify that 99 percent of the incidents were resolved within the time and cost frame, as agreed-on with the CSP.		

Figure 37—IT Audit and Assurance Program for Cloud Computing *(cont.)*

B—Understand Enablers, Set Suitable Assessment Criteria and Perform the Assessment *(cont.)*

Ref.	Assurance Steps and Guidance				Issue Cross-reference	Comment
B-3.4a *(cont.)*	Supplier relationships are working effectively.	Number of cloud supplier review meetings	On a regular basis, there are cloud supplier review meetings between the enterprise and the CSP to discuss performance of the cloud services.	Criteria: • Obtain the meeting minutes from all the cloud supplier meetings. • Verify that they were held on a regular basis. • Verify that action points from the previous meeting have been addressed, previous period cloud supplier performance was discussed and new action/improvement points are set forward.		
B-3.5a	Agree on suitable criteria to evaluate all processes in scope of the assurance engagement: define and agree on the reference process, i.e., determine which base practices a process should at least include. (This usually is just a confirmation of the COBIT 5 processes already identified, unless there is reason for using a different reference process.) Agree on the process practices that should be in place. (process design). Assess the **process design**, i.e., assess to what extent: • Expected process practices are applied. • Accountability and responsibility are assigned and assumed.					

Figure 37—IT Audit and Assurance Program for Cloud Computing *(cont.)*

B—Understand Enablers, Set Suitable Assessment Criteria and Perform the Assessment *(cont.)*					
Ref.	**Assurance Steps and Guidance**			**Issue Cross-reference**	**Comment**
B-3.5a *(cont.)*	The process APO10 *Manage suppliers* is in scope. The reference process is the COBIT 5 process with the same name as described in *COBIT 5: Enabling Processes.* The process required a number of management practices to be implemented, as described in the process description in the same guide. These are listed below, and they constitute: • A sound process design • The reference against which the process will be assessed with the criteria as mentioned, i.e., all management practices are expected to be fully implemented.		Each practice is typically implemented through a number of activities, and a well-designed process will implement these practices and activities.		
	Reference Process	APO10 Manage suppliers.	Criteria: All management practices fully implemented for CSPs.		
	Reference Process Practices	**Assessment Step**			
	APO10.01 Identify and evaluate cloud supplier relationships and contracts.	Assess by applying appropriate audit techniques (interview, observation, testing) whether the management practice is effectively implemented through the following typical (control) activities: 1. Establish and maintain criteria relating to type, significance and criticality of suppliers and supplier contracts, enabling a focus on preferred and important suppliers. 2. Establish and maintain supplier and contract evaluation criteria to enable overall review and comparison of supplier performance in a consistent way. 3. Identify, record and categorise existing suppliers and contracts according to defined criteria to maintain a detailed register of preferred suppliers that need to be managed carefully. 4. Periodically evaluate and compare the performance of existing and alternative suppliers to identify opportunities or a compelling need to reconsider current supplier contracts.			

Figure 37—IT Audit and Assurance Program for Cloud Computing *(cont.)*

Ref.	Assurance Steps and Guidance		Issue Cross-reference	Comment
B—Understand Enablers, Set Suitable Assessment Criteria and Perform the Assessment *(cont.)*				
B-3.5a *(cont.)*	**Reference Process Practices**	**Assessment Step**		
	APO10.01 *(cont.)*	Compare the RACI chart as included in the reference process in *COBIT 5: Enabling Processes* with the actual accountability and responsibility for this practice and assess whether: • Accountability and responsibility are assigned and assumed. • Accountability and responsibility are assigned at the appropriate level in the enterprise.		
	APO10.02 Select suppliers.	Assess by applying appropriate audit techniques (interview, observation, testing) whether emergency changes are effectively managed through the following typical (control) activities: 1. Review all requests for information (RFIs) and requests for proposal (RFPs) to ensure that they: • Clearly define requirements • Include a procedure to clarify requirements • Allow vendors sufficient time to prepare their proposals • Clearly define award criteria and the decision process 2. Evaluate RFIs and RFPs in accordance with the approved evaluation process/criteria, and maintain documentary evidence of the evaluations. Verify the references of candidate vendors. 3. Select the supplier that best fits the RFP. Document and communicate the decision, and sign the contract. 4. In the specific case of software acquisition, include and enforce the rights and obligations of all parties in the contractual terms. These rights and obligations may include ownership and licencing of intellectual property, maintenance, warranties, arbitration procedures, upgrade terms, and fit for purpose, including security, escrow and access rights.		

Figure 37—IT Audit and Assurance Program for Cloud Computing *(cont.)*

B—Understand Enablers, Set Suitable Assessment Criteria and Perform the Assessment *(cont.)*				
Ref.	**Assurance Steps and Guidance**		**Issue Cross-reference**	**Comment**
B-3.5a *(cont.)*	**Reference Process Practices**	**Assessment Step**		
	AP010.02 *(cont.)*	5. In the specific case of acquisition of development resources, include and enforce the rights and obligations of all parties in the contractual terms. These rights and obligations may include ownership and licencing of intellectual property; fit for purpose, including development methodologies; testing; quality management processes, including required performance criteria; performance reviews; basis for payment; warranties; arbitration procedures; human resource management; and compliance with the enterprise's policies. 6. Obtain legal advice on resource development acquisition agreements regarding ownership and licencing of intellectual property. 7. In the specific case of acquisition of infrastructure, facilities and related services, include and enforce the rights and obligations of all parties in the contractual terms. These rights and obligations may include service levels, maintenance procedures, access controls, security, performance review, basis for payment and arbitration procedures. Compare the RACI chart as included in the reference process in *COBIT 5: Enabling Processes* with the actual accountability and responsibility for this practice and assess whether: • Accountability and responsibility are assigned and assumed. • Accountability and responsibility are assigned at the appropriate level in the enterprise.		

Figure 37—IT Audit and Assurance Program for Cloud Computing *(cont.)*

B—Understand Enablers, Set Suitable Assessment Criteria and Perform the Assessment *(cont.)*

Ref.	Assurance Steps and Guidance		Issue Cross-reference	Comment
B-3.5a *(cont.)*	**Reference Process Practices**	**Assessment Step**		
	AP010.03 Manage supplier relationships and contracts.	Assess by applying appropriate audit techniques (interview, observation, testing) whether the management practice is effectively implemented through the following typical (control) activities: 1. Assign relationship owners for all suppliers and make them accountable for the quality of service(s) provided. 2. Specify a formal communication and review process, including supplier interactions and schedules. 3. Agree on, manage, maintain and renew formal contracts with the supplier. Ensure that contracts conform to enterprise standards and legal and regulatory requirements. 4. Within contracts with key service suppliers include provisions for the review of supplier site and internal practices and controls by management or independent third parties. 5. Evaluate the effectiveness of the relationship and identify necessary improvements. 6. Define, communicate and agree on ways to implement required improvements to the relationship. 7. Use established procedures to deal with contract disputes, first using, wherever possible, effective relationships and communications to overcome service problems. 8. Define and formalise roles and responsibilities for each service supplier. Where several suppliers combine to provide a service, consider allocating a lead contractor role to one of the suppliers to take responsibility for an overall contract. Compare the RACI chart as included in the reference process in *COBIT 5: Enabling Processes* with the actual accountability and responsibility for this practice and assess whether: • Accountability and responsibility are assigned and assumed. • Accountability and responsibility are assigned at the appropriate level in the enterprise.		

Figure 37—IT Audit and Assurance Program for Cloud Computing *(cont.)*

Ref.	Assurance Steps and Guidance		Issue Cross-reference	Comment
B-3.5a *(cont.)*	**Reference Process Practices**	**Assessment Step**		
	AP010.04 Manage supplier risk.	Assess by applying appropriate audit techniques (interview, observation, testing) whether changes are effectively closed and documented through the following typical (control activities: 1. Identify, monitor and, where appropriate, manage risk relating to the supplier's ability to deliver service efficiently, effectively, securely, reliably and continually. 2. When defining the contract, provide for potential service risk by clearly defining service requirements, including software escrow agreements, alternative suppliers or standby agreements to mitigate possible supplier failure; security and protection of IP; and any legal or regulatory requirements. Compare the RACI chart as included in the reference process in *COBIT 5: Enabling Processes* with the actual accountability and responsibility for this practice and assess whether: • Accountability and responsibility are assigned and assumed. • Accountability and responsibility are assigned at the appropriate level in the enterprise.		

Table section header: B—Understand Enablers, Set Suitable Assessment Criteria and Perform the Assessment *(cont.)*

Figure 37—IT Audit and Assurance Program for Cloud Computing *(cont.)*

B—Understand Enablers, Set Suitable Assessment Criteria and Perform the Assessment *(cont.)*				
Ref.	**Assurance Steps and Guidance**		**Issue Cross-reference**	**Comment**
B-3.5a *(cont.)*	**Reference Process Practices**	**Assessment Step**		
	APO10.05 Monitor supplier performance and compliance.	Assess by applying appropriate audit techniques (interview, observation, testing) whether changes are effectively closed and documented through the following typical (control activities: 1. Define and document criteria to monitor supplier performance aligned with SLAs and ensure that the supplier regularly and transparently reports on agreed-on criteria. 2. Monitor and review service delivery to ensure that the supplier is providing an acceptable quality of service, meeting requirements and adhering to contract conditions. 3. Review supplier performance and value for money to ensure that they are reliable and competitive, compared with alternative suppliers and market conditions. 4. Request independent reviews of supplier internal practices and controls, if necessary. 5. Record and assess review results periodically and discuss them with the supplier to identify needs and opportunities for improvement. 6. Monitor and evaluate externally available information about the supplier. Compare the RACI chart as included in the reference process in *COBIT 5: Enabling Processes* with the actual accountability and responsibility for this practice and assess whether: • Accountability and responsibility are assigned and assumed. • Accountability and responsibility are assigned at the appropriate level in the enterprise.		

Figure 37—IT Audit and Assurance Program for Cloud Computing *(cont.)*

Ref.	Assurance Steps and Guidance			Issue Cross-reference	Comment
B—Understand Enablers, Set Suitable Assessment Criteria and Perform the Assessment *(cont.)*					
B-3.6a	**Reference Process Practices**	**Assessment Step**			
	Agree on the **process work products** (inputs and outputs as defined in the process practices description) that are expected to be present (process design). Assess to what extent the process work products are available.				
	The process AP010 *Manage suppliers* identifies a set of inputs and outputs for the different management practices. The most relevant (and not assessed as Information items in scope in section A-3.3) of these work products are identified as follows, as well as the criteria to assess them against, i.e., existence and usage.		Criteria: All listed work products should demonstrably exist and be used.		
	Process Practice	**Work Products**	**Assessment Step**		
	AP010.01	Potential revision to CSP contracts	Apply appropriate audit techniques to determine for each work product: • Existence of the work product • Appropriate use of the work product		
	AP010.02	RFI and RFP evaluations			
	AP010.03	• Approved acquisition plans • Communication and review process • Review results and suggested improvements			
	AP010.04	• Cloud supplier risk analysis and risk profile reports for stakeholders			
	AP010.05	• CSP compliance monitoring criteria • CSP compliance monitoring review results			

Figure 37—IT Audit and Assurance Program for Cloud Computing *(cont.)*

B—Understand Enablers, Set Suitable Assessment Criteria and Perform the Assessment *(cont.)*			
Ref.	**Assurance Steps and Guidance**	**Issue Cross-reference**	**Comment**
B-3.7a	Agree on the **process capability level** to be achieved by the process. Assess the achievement of process capability level 1 Performed Process. Assess the achievement of higher process capability levels.		
	For a process capability level assessment, the ISO/IEC 15504:2 standard or the COBIT 5 PAM should be consulted. Process APO10 *Manage suppliers* is important and will require the following process capability level and attributes, which is equivalent to achieving a process capability level 2. (This step is only warranted if the process under review is a "standard" COBIT 5 governance or management process to which the ISO/IEC 15504 PAM can be applied. Any other processes, for which no reference practices, work products or outcomes are approved cannot use this assessment method, and hence the concept capability level does not apply.)		

Figure 37—IT Audit and Assurance Program for Cloud Computing *(cont.)*

B—Understand Enablers, Set Suitable Assessment Criteria and Perform the Assessment *(cont.)*			
Ref.	**Assurance Steps and Guidance**	**Issue Cross-reference**	**Comment**
B-4	**Obtain understanding of each Organisational Structure in scope and set suitable assessment criteria: for each Organisational Structure in scope (as determined in step A-3.1), additional information is collected and assessment criteria are defined. Assess the Organisational Structure.**		
Repeat Steps B-4.1 through B-4.5 for **all remaining Organisational Structures in scope**. a. CISO b. Business executives c. Service manager d. CIO e. Business process owners			
B-4.1a	Understand the Organisational Structure context.		
B-4.2a	Understand all stakeholders of the Organisational Structure/function.		

<table>
<tr><th colspan="5">Figure 37—IT Audit and Assurance Program for Cloud Computing (cont.)</th></tr>
<tr><th colspan="5">B—Understand Enablers, Set Suitable Assessment Criteria and Perform the Assessment (cont.)</th></tr>
<tr><th>Ref.</th><th colspan="2">Assurance Steps and Guidance</th><th>Issue Cross-reference</th><th>Comment</th></tr>
<tr><td rowspan="2">B-4.3a</td><td colspan="2"><u>Understand</u> the goals of the Organisational Structure, the related metrics and agree expected values. <u>Understand</u> how these goals contribute to the achievement of the enterprise goals and IT-related goals</td><td></td><td></td></tr>
<tr><td>Looking at COBIT 5: Enabling Processes, and considering the scope of the review, the CISO is primarily accountable for the following practices:
• APO13.01 Establish and maintain an ISMS.
• APO13.02 Define and manage an information security risk treatment plan.
• APO13.03 Monitor and review the ISMS.
• DSS05.02 Manage network and connectivity security.

All of these are already part of the process review, i.e., during the review of the process (as described in sections B-2.1x through B-2.7x) the existence and execution of the different process practices and activities is assessed, which provides an answer to the question whether the goals of the Organisational Structure/role are achieved. Hence, no specific additional assurance steps are required here.</td><td>For the Organisational Structure at hand—CISO—goals consist of a set of actions/decisions as described in B-3.3x. The performance and outcome of these actions/decisions are already part of the process review of the processes concerned (APO13 and DSS05). For purposes of completeness they are nonetheless listed below.

In addition, the goal of the Organisational Structures' decisions is to support the achievement of the IT-related and the enterprise goals. The assessment could therefore include another step to reflect on the assessment results of step C-1.1 and C-1.2, i.e., to consider to what extent the Organisational Structures have contributed to or have failed contributing to the achievement of these goals.</td><td></td><td></td></tr>
</table>

Figure 37—IT Audit and Assurance Program for Cloud Computing *(cont.)*

B—Understand Enablers, Set Suitable Assessment Criteria and Perform the Assessment *(cont.)*				
Ref.	**Assurance Steps and Guidance**		**Issue Cross-reference**	**Comment**
B-4.3a *(cont.)*	**Organisational Structure Goal**	**Assessment Step**		
	Contribution: Assess contribution to IT-related goal ITG02	• Evaluate whether this decision has positively contributed to the achievement of the stated IT goals. • Assess whether—in case the goal has not been achieved, to what extent the Organisational Structure has failed to contribute.		
	Contribution: Assess contribution to IT-related goal ITG10	• Evaluate whether this decision has positively contributed to the achievement of the stated IT goals. • Assess whether—in case the goal has not been achieved, to what extent the Organisational Structure has failed to contribute.		
	Contribution to enterprise goal EG03	• Evaluate whether this decision has positively contributed to the achievement of the stated enterprise goals. • Assess whether—in case the goal has not been achieved, to what extent the Organisational Structure has failed to contribute.		
	Contribution to enterprise goal EG04	• Evaluate whether this decision has positively contributed to the achievement of the stated enterprise goals. • Assess whether—in case the goal has not been achieved, to what extent the Organisational Structure has failed to contribute.		

Figure 37—IT Audit and Assurance Program for Cloud Computing *(cont.)*

Ref.	Assurance Steps and Guidance			Issue Cross-reference	Comment
B—Understand Enablers, Set Suitable Assessment Criteria and Perform the Assessment *(cont.)*					
B-4.4a	Agree on the expected good practices for the Organisational Structure against which it will be assessed. Assess the **Organisational Structure design**, i.e., assess to what extent expected **good practices** are applied.				
	Expected good practice and assessment criteria for the CISO include the following:				
	Good Practice	**Criteria**	**Assessment Step**		
	Operating principles	• Operating principles are documented. • Action and achievement reports are available and are meaningful.	• Verify whether operating principles are appropriately documented. • Verify that action and achievement reports are available and are meaningful.		
	Span of control	• The span of control of the Organisational Structure is defined. • The span of control is adequate, i.e., the Organisational Structure has the right to take all decisions it should. • The span of control is in line with the overall enterprise governance arrangements.	• Verify whether the span of control of the Organisational Structure is defined. • Assess whether the span of control is adequate, i.e., the Organisational Structure has the right to take all decisions it should. • Verify and assess whether the span of control is in line with the overall enterprise governance arrangements.		

Figure 37—IT Audit and Assurance Program for Cloud Computing *(cont.)*

B—Understand Enablers, Set Suitable Assessment Criteria and Perform the Assessment *(cont.)*					
Ref.	**Assurance Steps and Guidance**			**Issue Cross-reference**	**Comment**
B-4.4a *(cont.)*	**Good Practice**	**Criteria**	**Assessment Step**		
	Level of authority/decision rights	• Decision rights of the Organisational Structure are defined and documented. • Decision rights of the Organisational Structure are respected and complied with (also a culture/behaviour issue).	• Verify that decision rights of the Organisational Structure are defined and documented. • Verify whether decision rights of the Organisational Structure are complied with and respected.		
	Delegation of authority	Delegation of authority is implemented in a meaningful way.	Verify whether delegation of authority is implemented in a meaningful way.		
	Escalation procedures	There are escalation procedures defined and applied.	Verify the existence and application of escalation procedures.		

Figure 37—IT Audit and Assurance Program for Cloud Computing *(cont.)*

Ref.	Assurance Steps and Guidance			Issue Cross-reference	Comment
B—Understand Enablers, Set Suitable Assessment Criteria and Perform the Assessment *(cont.)*					
B-4.5a	Understand the life cycle and agree on expected values. Assess the extent to which the **Organisational Structure life cycle** is managed.				
	The following criteria relate to life cycle management should be complied with by the CISO.		For the CISO, assess by applying appropriate auditing techniques the following life cycle management elements.		
	Life Cycle Element	**Criteria**	**Assessment Step**		
	Mandate	• The Organisational Structure is formally established. • The Organisational Structure has a clear, documented and well understood mandate.	• Verify through interviews and observations that the Organisational Structure is formally established. • Verify through interviews and observations that the Organisational Structure has a clear, documented and well understood mandate.		
	Monitoring	• The performance of the Organisational Structure should be regularly monitored and evaluated by competent and independent assessors. • The regular evaluations should result in the required continuous improvements to the Organisational Structure, either in its span of control, level of authority or any other parameter.	• Verify whether the performance of the Organisational Structure is regularly monitored and evaluated by competent and independent assessors. • Verify whether the regular evaluations have resulted in improvements to the Organisational Structure, either in its span of control, level of authority or any other parameter.		

Figure 37—IT Audit and Assurance Program for Cloud Computing *(cont.)*

B—Understand Enablers, Set Suitable Assessment Criteria and Perform the Assessment *(cont.)*

Ref.	Assurance Steps and Guidance	Issue Cross-reference	Comment
B-5	Obtain understanding of the **Culture, Ethics and Behaviour** in scope. Assess Culture, Ethics and Behaviour.		
Repeat Steps B-5.1 through B-5.5 for all remaining **Culture, Ethics and Behaviour aspects in scope**: vendor collaboration is a key success factor: a. **Data confidentiality is enterprisewide recognized as an attention point when using cloud computing services**. b. Management proactively monitors risk and action plan progress. c. Policies reflect risk appetite and risk tolerance. d. The value of risk is recognized.			
B-5.1a	Understand the Culture, Ethics and Behaviour context.		
B-5.2a	Understand the major stakeholders of the Culture, Ethics and Behaviour.		
	The behaviour at hand—Data confidentiality is enterprisewide recognized as an attention point when using cloud computing services, and it will apply to all those roles and individuals in the enterprise who deal with acquiring and using cloud computing services.		

Figure 37—IT Audit and Assurance Program for Cloud Computing *(cont.)*

B—Understand Enablers, Set Suitable Assessment Criteria and Perform the Assessment *(cont.)*				
Ref.	**Assurance Steps and Guidance**		**Issue Cross-reference**	**Comment**
B-5.3a	Understand the goals for the Culture, Ethics and Behaviour, the related metrics and agree on expected values. Assess whether the **Culture, Ethics and Behaviour goals** (outcomes) are achieved, i.e., assess the effectiveness of the Culture, Ethics and Behaviour.			
	In the context of this assurance engagement and the behaviour at hand, the desired behaviour follows.			
	Desired Behaviour (Culture, Ethics and Behaviour Goal)	**Assessment Step**		
	Employees using cloud services are aware that all of the data is transmitted through the network to an offsite location.	Verify that awareness campaigns have been conducted, to raise awareness on the specifics of using cloud computing services.		
	When setting up contracts with a CSP, management involved should be aware of the security and privacy risk that is being faced and should foresee appropriate requirements in the contract/SLA.	• Obtain the contracts and SLAs with the enterprise's CSPs. • Verify that requirements and mitigating actions have been set up in the contracts and SLAs with the CSPs with regard to ensuring data privacy and confidentiality.		
B-5.4a	Understand the life cycle stages of the Culture, Ethics and Behaviour, and agree on the relevant criteria. Assess to what extent the Culture, Ethics and Behaviour life cycle is managed.			
	(This aspect is already covered by the assessment of the good practices, hence no additional separate assurance steps are defined here.)			

Figure 37—IT Audit and Assurance Program for Cloud Computing *(cont.)*

B—Understand Enablers, Set Suitable Assessment Criteria and Perform the Assessment *(cont.)*					
Ref.	**Assurance Steps and Guidance**			**Issue Cross-reference**	**Comment**
B-5.5a	Understand good practice when dealing with Culture, Ethics and Behaviour, and agree on relevant criteria. Assess the Culture, Ethics And Behaviour design, i.e., assess to what extent expected good practices are applied.				
	In the context of this assurance engagement and the behavior at hand, it is expected that the following good practices are all applied.				
	Good Practice	**Criteria**	**Assessment Step**		
	Communication, enforcement and rules	Specific rules for minimal security of data transmitted over the network to a CSP are set up and should be (over) achieved by every contracted CSP.	Verify that a document exists that list the minimal security requirements for both the network as the infrastructure facilities when using cloud services.		
	Incentives and rewards	Provide incentives for the CSP to continue to strive for optimal data confidentiality.	Verify that the SLAs with the CSPs provide incentives and penalties on data confidentiality for the CSP to continue to strive for optimal data confidentiality.		
	Awareness	Employees are being informed about the specificities around data privacy of using cloud computing services.	Verify that awareness campaigns have been conducted, to raise awareness on the specifics of using cloud computing services.		

Figure 37—IT Audit and Assurance Program for Cloud Computing *(cont.)*

Ref.	Assurance Steps and Guidance	Issue Cross-reference	Comment
B—Understand Enablers, Set Suitable Assessment Criteria and Perform the Assessment *(cont.)*			
B-6	Obtain understanding of the **Information items** in scope. Assess Information items.		
Repeat Steps B-6.1 through B-6.5 for all remaining Information items in scope. • **Evaluations against SLAs** • IT-related policies (output, APO01.03; input, all APO, all BAI, all DSS, all MEA) • Communication on IT objectives (output APO01.04; input, all APO, all BAI, all DSS, all MEA) • Data classification guidelines (output, APO01.06; input APO03.02, BAI02.01, DSS05.02, DSS06.01) • Data security and control guidelines (output, APO01.06; input, BAI02.01) • Data integrity procedures (output, APO01.06; input, BAI02.01, DSS06.01) • Non-compliance remedial actions (output, APO01.08; input, MEA01.05) • SLAs (output, APO09.03) • Service level performance reports (output, APO09.04) • Supplier contracts (input, APO10.01) • Identified supplier delivery risk (output, APO10.04; input, APO12.01, APO12.03, BAI01.01) • Identified contract requirements to minimise risk (output, APO10.04) • Risk analysis results (output APO12.02; input EDM03.03, APO01.03, APO02.02, BAI01.10) • Review results of third-party risk assessments (output, APO12.04; input, EDM03.03, APO10.04, MEA02.01) • Results of penetration tests (output, DSS05.02; input, MEA02.08) • Identified compliance gaps (output, MEA03.03; input, MEA02.08) • Reports of noncompliance issues and root causes (output, MEA03.04; input, EDM01.03, MEA02.07)			
B-6.1a	Understand the Information item context.		
B-6.2a	Understand the major stakeholders of the Information item.		

Figure 37—IT Audit and Assurance Program for Cloud Computing *(cont.)*

B—Understand Enablers, Set Suitable Assessment Criteria and Perform the Assessment *(cont.)*

Ref.	Assurance Steps and Guidance				Issue Cross-reference	Comment
B-6.3a	Understand the major quality criteria for the Information item, the related metrics and agree expected values. Assess whether the **Information item quality criteria** (outcomes) are achieved, i.e., assess the effectiveness of the Information item.					
	When looking at the Information item at hand, evaluations against SLAs, the quality dimensions marked with a ✔ are deemed most important, and by consequence will be assessed against the described criteria.			For the information item at hand—evaluations against SLAs—the assurance professional will perform the following assurance steps.		
	Quality Dimension	**Key Criteria**	**Description**	**Assessment Step**		
	Accuracy	✔	The evaluation should give an accurate description of the established service levels, the evaluation performed against the service levels and the data used for the evaluation.	Verify based on a sample of performed evaluations that these provide an accurate description of the evaluation itself, data used for the evaluation and the service levels against which the evaluation has been performed.		
	Objectivity	✔	The evaluation should be based on identified and objective performance indicators.	Verify based on a sample that the evaluations were performed based on gathered objective performance data.		

Figure 37—IT Audit and Assurance Program for Cloud Computing *(cont.)*

B—Understand Enablers, Set Suitable Assessment Criteria and Perform the Assessment *(cont.)*						
Ref.	**Assurance Steps and Guidance**				**Issue Cross-reference**	**Comment**
B-6.3a *(cont.)*	**Quality Dimension**	**Key Criteria**	**Description**	**Assessment Step**		
	Believability					
	Reputation					
	Relevancy					
	Completeness	✔	The evaluation covers the complete set of agreed-on service levels with the CSP.	Verify based on a sample that the evaluations cover the entire set of agreed-on service levels with the CSP, and does not solely evaluate the performance on the most important service levels.		
	Currency	✔	The evaluation is based on data that is current to the SLA validity period under review.	Verify that the time frame of the data being used, is concurrent with the validity period of the SLA.		
	Amount of information					
	Concise representation					
	Consistent representation					
	Interpretability					

Figure 37—IT Audit and Assurance Program for Cloud Computing *(cont.)*

B—Understand Enablers, Set Suitable Assessment Criteria and Perform the Assessment *(cont.)*						
Ref.	**Assurance Steps and Guidance**				**Issue Cross-reference**	**Comment**
B-6.3a *(cont.)*	**Quality Dimension**	**Key Criteria**	**Description**	**Assessment Step**		
	Understandability	✓	The evaluation and subsequent conclusions should be clear to all stakeholders for them to take relevant actions.	Verify based on a sample that the evaluations are clear to all stakeholders.		
	Manipulation					
	Availability					
	Restricted access					
B-6.4	<u>Understand</u> the life cycle stages of the Information item, and agree the relevant criteria. Assess to what extent the **Information item life cycle** is managed.					
	The Information item, evaluations against SLAs,, is an output of the process BAI04 *Manage availability* and capacity and its life cycle is covered almost entirely by that process, including the monitor and adjusting stages. BAI04 is not part of our key processes. As a result, specific assurance steps are required for some life cycle stages.		For the information item at hand—evaluations against SLAs—the assurance professional will perform the following assurance steps.			

Figure 37—IT Audit and Assurance Program for Cloud Computing *(cont.)*

B—Understand Enablers, Set Suitable Assessment Criteria and Perform the Assessment *(cont.)*						
Ref.	**Assurance Steps and Guidance**				**Issue Cross-reference**	**Comment**
B-6.4 *(cont.)*	**Life Cycle Stage**	**Key Criteria**	**Description**	**Assessment Step**		
	Plan	✔	The evaluation should be designed in such a way that it will evaluate the performance of the CSP on the agreed-on service levels in an efficient and effective way.	Verify that the evaluation was designed to ensure the evaluation of the service level in a correct, efficient and effective way.		
	Design					
	Build/acquire					
	Use/operate					
	Evaluate/monitor					
	Update/dispose	✔	The evaluation should be performed on a regular basis.	Verify that the evaluations are performed on a regular basis.		

<table>
<tr><th colspan="7">Figure 37—IT Audit and Assurance Program for Cloud Computing (cont.)</th></tr>
<tr><th colspan="7">B—Understand Enablers, Set Suitable Assessment Criteria and Perform the Assessment (cont.)</th></tr>
<tr><th>Ref.</th><th colspan="4">Assurance Steps and Guidance</th><th>Issue Cross-reference</th><th>Comment</th></tr>
<tr><td>B-6.5a</td><td colspan="4"><u>Understand</u> important attributes of the Information item and expected values.
<u>Assess</u> the Information item design, i.e., assess to what extent expected good practices are applied.</td><td></td><td></td></tr>
<tr><td></td><td colspan="2">When looking at the Information item at hand, Evaluations against SLA's, some of the information attributes are relevant to be assessed, as follows.</td><td colspan="2"></td><td></td><td></td></tr>
<tr><td></td><th>Attribute</th><th>Key Criteria</th><th>Description</th><th>Assessment Step</th><td></td><td></td></tr>
<tr><td></td><td>Physical</td><td></td><td></td><td></td><td></td><td></td></tr>
<tr><td></td><td>Empirical</td><td></td><td></td><td></td><td></td><td></td></tr>
<tr><td></td><td>Syntactic</td><td>✔</td><td>The evaluation should be based on current information and provide sufficient for the reader to understand the conclusion.</td><td>Verify based on a sample that the evaluations are based on current information and provide sufficient details for the reader to understand the conclusion.</td><td></td><td></td></tr>
<tr><td></td><td>Semantic</td><td></td><td></td><td></td><td></td><td></td></tr>
<tr><td></td><td>Pragmatic</td><td></td><td></td><td></td><td></td><td></td></tr>
<tr><td></td><td>Social</td><td></td><td></td><td></td><td></td><td></td></tr>
</table>

<table>
<tr><th colspan="6">Figure 37—IT Audit and Assurance Program for Cloud Computing (cont.)</th></tr>
<tr><th colspan="6">B—Understand Enablers, Set Suitable Assessment Criteria and Perform the Assessment (cont.)</th></tr>
<tr><th>Ref.</th><th colspan="3">Assurance Steps and Guidance</th><th>Issue Cross-reference</th><th>Comment</th></tr>
<tr><td>B-7</td><td colspan="3">Obtain understanding of the Services, Infrastructure and Applications in scope.
Assess Services, Infrastructure and Applications.</td><td></td><td></td></tr>
<tr><td colspan="4"><u>Repeat</u> Steps B-7.1 through B-7.5 for all remaining Services, Infrastructure and Applications in scope.
a. <u>SLA monitoring tool</u>
b. Risk management tool
c. Service provider evaluation tool</td><td></td><td></td></tr>
<tr><td>B-7.1a</td><td colspan="3"><u>Understand</u> the Services, Infrastructure and Applications context.</td><td></td><td></td></tr>
<tr><td>B-7.2a</td><td colspan="3"><u>Understand</u> the major stakeholders of the Services, Infrastructure and Applications.</td><td></td><td></td></tr>
<tr><td rowspan="6">B-7.3a</td><td colspan="3"><u>Understand</u> the major goals for the Services, Infrastructure and Applications, the related metrics and agree expected values. Assess whether the Services, Infrastructure and Applications goals (outcomes) are achieved, i.e., assess the effectiveness of the Services, Infrastructure and Applications.</td><td></td><td></td></tr>
<tr><td colspan="3">In the context of this assignment and the service at hand, it is expected that the following goals be achieved.</td><td></td><td></td></tr>
<tr><th>Goals</th><th>Criteria</th><th>Assessment Step</th><td></td><td></td></tr>
<tr><td>Service definition</td><td>• The SLA monitoring tool is clearly defined.
• The SLA monitoring tool is available to all potential stakeholders.</td><td>• Verify that the SLA monitoring tool is clearly defined.
• Assess the quality of the tool and of the service offered.
• Verify the accessibility of the tool to all potential stakeholders.</td><td></td><td></td></tr>
<tr><td>Service level</td><td>Service levels are defined using:
• Quality of the tool
• Cost
• Timeliness</td><td>Verify that the following aspects are dealt with in the service level definitions:
• Quality of the tool
• Cost
• Timeliness</td><td></td><td></td></tr>
<tr><td>Contribution to enabler, IT-related and enterprise goals</td><td>The SLA monitoring tool contributes to the achievement of related enabler and IT-related and enterprise goals.</td><td>Assess to what extent the SLA monitoring tool and service contributes to the achievement of related IT goals and enterprise goals.</td><td></td><td></td></tr>
</table>

Figure 37—IT Audit and Assurance Program for Cloud Computing *(cont.)*

<table>
<tr><th colspan="6">B—Understand Enablers, Set Suitable Assessment Criteria and Perform the Assessment (cont.)</th></tr>
<tr><th>Ref.</th><th colspan="3">Assurance Steps and Guidance</th><th>Issue Cross-reference</th><th>Comment</th></tr>
<tr><td>B-7.4a</td><td colspan="3"><u>Understand</u> the life cycle stages of the Services, Infrastructure and Applications, and agree the relevant criteria.
<u>Assess</u> to what extent the Services, Infrastructure and Applications life cycle is managed.</td><td></td><td></td></tr>
<tr><td></td><td colspan="3">In the context of this assignment, and depending on the sourcing choices made, the processes APO09 and APO10 may be relevant.

Should the actual scoping of the assurance engagement indicate that these processes are essential, the processes may have to be added to the scope of the assurance engagement. Otherwise, the remainder of this assurance step assumes that it is sufficient to include achievement of goals and application of the good practices in the scope.</td><td></td><td></td></tr>
<tr><td>B-7.5a</td><td colspan="3"><u>Understand</u> good practice related to the Services, Infrastructure and Applications and expected values.
<u>Assess</u> the Services, Infrastructure and Applications design, i.e., assess to what extent expected good practices are applied.</td><td></td><td></td></tr>
<tr><td></td><th>Good Practice</th><th>Criteria</th><th>Assessment Step</th><td></td><td></td></tr>
<tr><td></td><td>Sourcing (buy/build)</td><td>A formal decision—based on a business case—needs to be taken regarding the sourcing of the tool.</td><td>• Verify that a formal decision—based on a business case—was taken regarding the sourcing of the tool.
• Verify the validity and quality of the business case.
• Verify that the sourcing decision has been duly executed.</td><td></td><td></td></tr>
<tr><td></td><td>Use</td><td>The use of the tool needs to be clear, i.e.:
• When it needs to be used and by whom
• The required compliance levels of the tool's output</td><td>• Verify that the use of the tool is clear, i.e., it is known when and by whom the tool needs to be used.
• Verify that actual use is in line with previous requirement.
• Verify that the actual tool's output is used adequately.</td><td></td><td></td></tr>
</table>

Figure 37—IT Audit and Assurance Program for Cloud Computing *(cont.)*

B—Understand Enablers, Set Suitable Assessment Criteria and Perform the Assessment *(cont.)*			
Ref.	**Assurance Steps and Guidance**	**Issue Cross-reference**	**Comment**
B-8	Obtain understanding of the **People, Skills and Competencies** in scope. Assess People, Skills and Competencies.		
Repeat Steps B-8.1 through B-8.5 for **all remaining People, Skills and Competencies in scope**. a. **Cloud Vendor Management Skills** b. Understanding of cloud security and compliance components c. Data integration skills			
B-8.1a	Understand the People, Skills and Competencies context.		
B-8.2a	Understand the major stakeholders for the People, Skills and Competencies.		
	The people, skills and competencies at hand, i.e., cloud vendor management skills, is an essential skill for all those involved in t selecting, negotiating with and managing vendors. In the context and scope of this assurance initiative, this relates to the following stakeholders, which all have Accountability in the APO processes in scope: • Service manager • CIO All these stakeholders and members of the Organisational Structure listed will need this skill to be effective when approving selecting, negotiating with and managing vendors.		

Figure 37—IT Audit and Assurance Program for Cloud Computing *(cont.)*

B—Understand Enablers, Set Suitable Assessment Criteria and Perform the Assessment *(cont.)*

<table>
<tr><th>Ref.</th><th colspan="3">Assurance Steps and Guidance</th><th>Issue Cross-reference</th><th>Comment</th></tr>
<tr><td rowspan="9">B-8.3a</td><td colspan="3">Understand the major goals for the People, Skills and Competencies, the related metrics and agree on expected values. Assess whether the People, Skills and Competencies goals (outcomes) are achieved, i.e., assess the effectiveness of the People, Skills and Competencies.</td><td></td><td></td></tr>
<tr><th>Goal</th><th>Criteria</th><th>Assessment Step</th><td></td><td></td></tr>
<tr><td>Experience</td><td>Average minimum level of four years relevant professional experience.</td><td>Verify that there is—across all individuals—an average minimum level of four years relevant professional experience.</td><td></td><td></td></tr>
<tr><td>Education</td><td>N/A</td><td>N/A</td><td></td><td></td></tr>
<tr><td>Qualification</td><td>N/A</td><td>N/A</td><td></td><td></td></tr>
<tr><td>Knowledge</td><td>Relevant knowledge includes knowledge about:
• Cloud IT costing/value
• Cloud IT architecture
• Cloud risk management and a cloud risk management framework</td><td>Verify through interviews and assessment of career overviews and curriculum vitae (CVs) that individuals have relevant knowledge about:
• Cloud IT costing/value
• Cloud IT architecture
• Cloud risk management and a cloud risk management framework</td><td></td><td></td></tr>
<tr><td>Technical skills</td><td>Basic cloud computing concepts</td><td>Verify through interviews and assessment of career overviews and CVs that individuals have relevant knowledge about cloud computing concepts.</td><td></td><td></td></tr>
<tr><td>Behavioural skills</td><td>• Communication—oral, presentation and written
• Negotiation skills</td><td>Verify through interviews, observations and surveys that individuals have adequate
• Communication—oral, presentation and written—skills
• Negotiation skills</td><td></td><td></td></tr>
<tr><td>Quantity of people with appropriate skill level</td><td>Appropriate mix in number of people with required skill level</td><td>Verify through interviews and assessment of career overviews and CVs that the quantity and distribution of people with required skill levels is appropriate across the whole cloud vendor management process.</td><td></td><td></td></tr>
</table>

Figure 37—IT Audit and Assurance Program for Cloud Computing *(cont.)*

B—Understand Enablers, Set Suitable Assessment Criteria and Perform the Assessment *(cont.)*

Ref.	Assurance Steps and Guidance			Issue Cross-reference	Comment
B-8.4a	Understand the life cycle stages of the **People, Skills and Competencies**, and agree the relevant criteria. Assess to what extent the People, Skills and Competencies life cycle is managed.				
	For the People, Skills and Competencies at hand, i.e., cloud vendor management skills, the life cycle phases and associated criteria can be expressed in function of the process APO07.		For the People, Skills and Competencies at hand—cloud vendor management skills—the assurance professional will perform the following assurance steps.		
	Life Cycle Element	**Criteria**	**Assessment Step**		
	Plan	Process APO07.03 activity 1 (Define the required and currently available skills and competencies of internal and external resources to achieve enterprise, IT and process goals.) is implemented in relation to this skill.	Assess whether process APO07.03 activity 1 is implemented in relation to this skill.		
	Design	Process APO07.03 activity 2 (Provide formal career planning and professional development to encourage competency development, opportunities for personal advancement and reduced dependence on key individuals.) and APO07.03 activity 3 (Provide access to knowledge repositories to support the development of skills and competencies.) are implemented in relation to this skill.	Assess whether process APO07.03 activity 2 and AF007.03 activity 3 are implemented in relation to this skill.		

Figure 37—IT Audit and Assurance Program for Cloud Computing *(cont.)*

B—Understand Enablers, Set Suitable Assessment Criteria and Perform the Assessment *(cont.)*

Ref.	Assurance Steps and Guidance			Issue Cross-reference	Comment
B-8.4a *(cont.)*	**Life Cycle Element**	**Criteria**	**Assessment Step**		
	Build	Process AP007.03 activity 4 (Identify gaps between required and available skills and develop action plans to address them on an individual and collective basis, such as training [technical and behavioural skills], recruitment, redeployment and changed sourcing strategies.) is implemented in relation to this skill.	Assess whether process AP007.03 activity 4 is implemented in relation to this skill.		
	Operate	Process AP007.03 activity 5 (Develop and deliver training programmes based on organisational and process requirements, including requirements for enterprise knowledge, internal control, ethical conduct and security.) is implemented in relation to this skill.	Assess whether process AP007.03 activity 5 is implemented in relation to this skill.		

Figure 37—IT Audit and Assurance Program for Cloud Computing *(cont.)*

B—Understand Enablers, Set Suitable Assessment Criteria and Perform the Assessment *(cont.)*					
Ref.	**Assurance Steps and Guidance**			**Issue Cross-reference**	**Comment**
B-8.4a *(cont.)*	**Life Cycle Element**	**Criteria**	**Assessment Step**		
	Evaluate	Process AP007.03 activity 6 (Conduct regular reviews to assess the evolution of the skills and competencies of the internal and external resources. Review succession planning.) is implemented in relation to this skill.	Assess whether process AP007.03 activity 6 is implemented in relation to this skill.		
	Update/dispose	Process AP007.03 activity 7 (Review training materials and programmes on a regular basis to ensure adequacy with respect to changing enterprise requirements and their impact on necessary knowledge, skills and abilities.) is implemented in relation to this skill.	Assess whether process AP007.03 activity 7 is implemented in relation to this skill.		

Figure 37—IT Audit and Assurance Program for Cloud Computing *(cont.)*

Ref.	Assurance Steps and Guidance		Issue Cross-reference	Comment
	B—Understand Enablers, Set Suitable Assessment Criteria and Perform the Assessment *(cont.)*			
B-8.5a	Understand good practice related to the **People, Skills and Competencies** and expected values. Assess the People, Skills and Competencies design, i.e., assess to what extent expected good practices are applied.			
	Good Practice	**Assessment Step**		
	Skill set and competencies are defined.	• Determine that an inventory of Skills and Competencies is maintained by organizational unit, job function and individual. • Evaluate the relevance and the contribution of the Skills and Competencies to the achievement of the goals of the Organisational Structure, and by consequence, IT-related goals and enterprise goals. • Evaluate the gap analysis between necessary portfolio of Skills and Competencies and current inventory of Skills and capabilities.		
	Skill levels are defined.	• Assess the flexibility and performance of meeting Skills development to address identified gaps between necessary and current Skill levels. • Assess the process for 360° performance evaluations.		

Figure 37—IT Audit and Assurance Program for Cloud Computing *(cont.)*

C—Communication		
Ref.	**Assurance Step**	**Guidance**
C-1	**Document exceptions and gaps.**	
C-1.1	Understand and document weaknesses and their impact on the achievement of process goals.	• Illustrate the impact of enabler failures or weaknesses with numbers and scenarios of errors, inefficiencies and misuse. • Clarify vulnerabilities, threats and missed opportunities that are likely to occur if enablers do not perform effectively.
C-1.2	Understand and document weaknesses and their impact on enterprise goals.	• Illustrate what the weaknesses would affect (e.g., business goals and objectives, enterprise architecture elements, capabilities, resources). Relate the impact of not achieving the enabler goals to actual cases in the same industry and leverage industry benchmarks. • Document the impact of actual enabler weaknesses in terms of bottom-line impact, integrity of financial reporting, hours lost in staff time, loss of sales, ability to manage and react to the market, customer and shareholder requirements, etc. • Point out the consequence of non-compliance with regulatory requirements and contractual agreements. • Measure the actual impact of disruptions and outages on business processes and objectives, and on customers (e.g., number, effort, downtime, customer satisfaction, cost).

Figure 37—IT Audit and Assurance Program for Cloud Computing *(cont.)*

C—Communication *(cont.)*		
Ref.	**Assurance Step**	**Guidance**
C-2	**Communicate.**	
C-2.1	Communicate the work performed.	• Communicate regularly to the stakeholders identified in A-1 on progress of the work performed.
C-2.2	Communicate preliminary findings to the assurance engagement stakeholders defined in A-1.	• Document the impact (i.e., customer and financial impact) of errors that could have been caught by effective enablers. • Measure and document the impact of rework (e.g., ratio of rework to normal work) as an efficiency measure affected by enabler weaknesses. • Measure the actual business benefits and illustrate cost savings of effective enablers after the fact. • Use benchmarking and survey results to compare the enterprise's performance with others. • Use extensive graphics to illustrate the issues. • Inform the person responsible for the assurance activity about the preliminary findings and verify his/her correct understanding of those findings.
C-2.3	Deliver a report (aligned with the terms of reference, scope and agreed-on reporting standards) that supports the results of the initiative and enables a clear focus on key issues and important actions.	

Page intentionally left blank

Appendix C. Process Capability Assessment

One of the consistent requests of stakeholders who have undergone IT audit/assurance reviews is a desire to understand how their performances compare to good practices. Audit and assurance professionals must provide an objective basis for the review conclusions. This chapter provides an overview of the COBIT 5 capability assessment model that can be used to assess the cloud-computing processes and determine their capability level. Examples of cloud computing-specific enablers are also provided for each of the capability levels, to increase the understanding of the practical applicability of the capability model.

Capability Assessment and the Assurance Initiative

The capability assessment is one of the final steps in the assurance initiative. The IT audit and assurance professional can address the key enablers within the scope of the work program and formulate an objective assessment of the capability levels of the enablers. The capability assessment can be a part of the audit/assurance report and can be used as a metric from year to year to document progression in the enhancement of the processes. However, it must be noted that the perception of the capability level may vary between the process/IT asset owner and the IT audit and assurance professional. Therefore, an IT audit and assurance professional should obtain the concerned stakeholder's concurrence before submitting the final report to management.

COBIT 5 Capability Assessment Based on ISO/IEC 15504

The COBIT 5 product set[33] includes a process capability model, based on the internationally recognized ISO/IEC 15504 Software Engineering—Process Assessment standard. This model will achieve the same overall objectives of process assessment and process improvement support, i.e., it will provide a means to measure the performance of any of the governance (EDM-based) processes or management (PBRM-based) processes and will allow areas for improvement to be identified.

The COBIT 5 ISO/IEC 15504-based assessment approach also facilitates the following objectives:

- Enable the governance body and management to benchmark process capability
- Enable high-level "as-is" and "to-be" health checks to support the governance body and management investment decision making with regard to the process improvement
- Provide gap analysis and improvement planning information to support definition of justifiable improvement projects
- Provide the governance body and management with assessment ratings to measure and monitor current capabilities

The COBIT 5 process capability approach can be summarized as shown in **figure 38**.

[33] ISACA, COBIT 5, USA, 2012, p. 42-45

Figure 38—COBIT 5 Process Capability Model

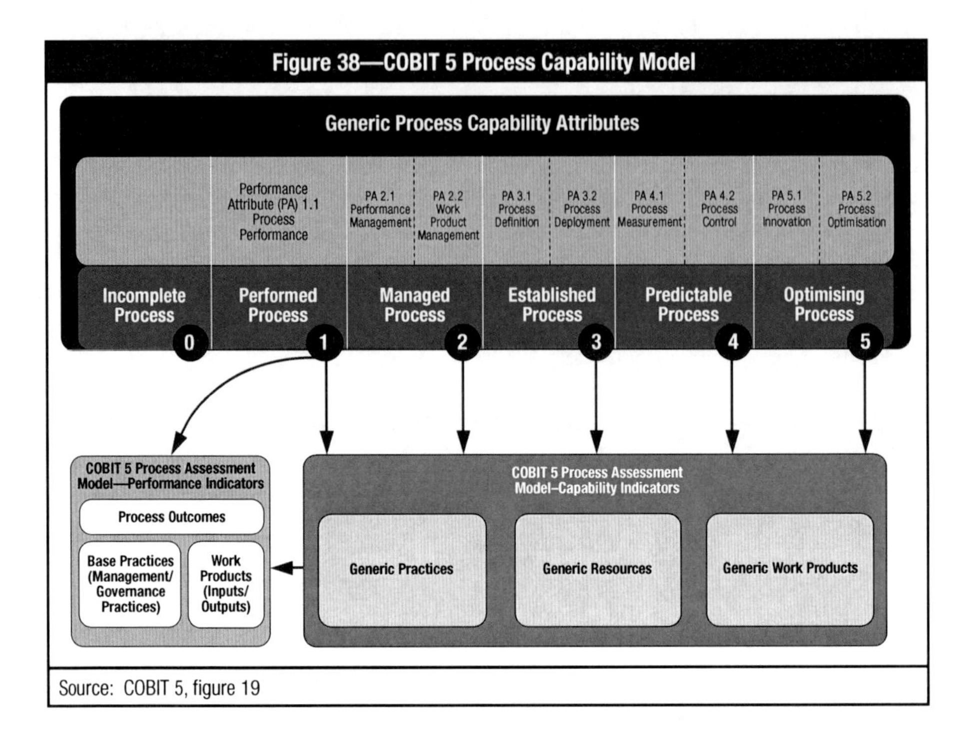

Source: COBIT 5, figure 19

There are six levels of capability that a process can achieve, including an "incomplete process" designation if the practices do not achieve the intended purpose of the process:

0 **Incomplete process**—The process is not implemented or fails to achieve its process purpose. At this level, there is little or no evidence of any systematic achievement of the process purpose.

1 **Performed process** (one attribute)—The implemented process achieves its process purpose.

2 **Managed process** (two attributes)—The previously described performed process is now implemented in a managed fashion (planned, monitored and adjusted) and its work products are appropriately established, controlled and maintained.

3 **Established process** (two attributes)—The previously described managed process is now implemented using a defined process that is capable of achieving its process outcomes.

4 **Predictable process** (two attributes)—The previously described established process now operates within defined limits to achieve its process outcomes.

5 **Optimising process** (two attributes)—The previously described predictable process is continuously improved to meet relevant, current and projected business goals.

Each capability level can be achieved only when the level below it has been fully achieved. For example, a process capability level 3 (established process) requires the process definition and process deployment attributes to be largely achieved, on top of full achievement of the attributes for a process capability level 2 (managed process).

There is a significant distinction between process capability level 1 and the higher capability levels. Process capability level 1 achievement requires the process performance attribute to be largely achieved, which actually means that the process is being successfully performed and the required outcomes are obtained by the enterprise. The higher capability levels then add different attributes to the process. In this assessment scheme, achieving a capability level 1, even on a scale to 5, is already an important achievement for an enterprise. Note that each individual enterprise shall choose (based on cost-benefit and feasibility reasons) its target or desired level, which very seldom will be one of the highest.

COBIT 5 Capability Assessment for Cloud Computing

Based on the previously explained COBIT 5 process capability model, this section provides examples that show how some of the primary elements of the enablers will look for each of the capability levels of the model. This allows for a more practical guidance on assessing the process capability level of the cloud computing processes.

Stepped Approach to the Assessment

The example follows the stepped approach, as shown in **figure 39** and explained in the *COBIT Self-Assessment Guide: Using COBIT 5.*

Figure 39—Stepped Assessment Approach

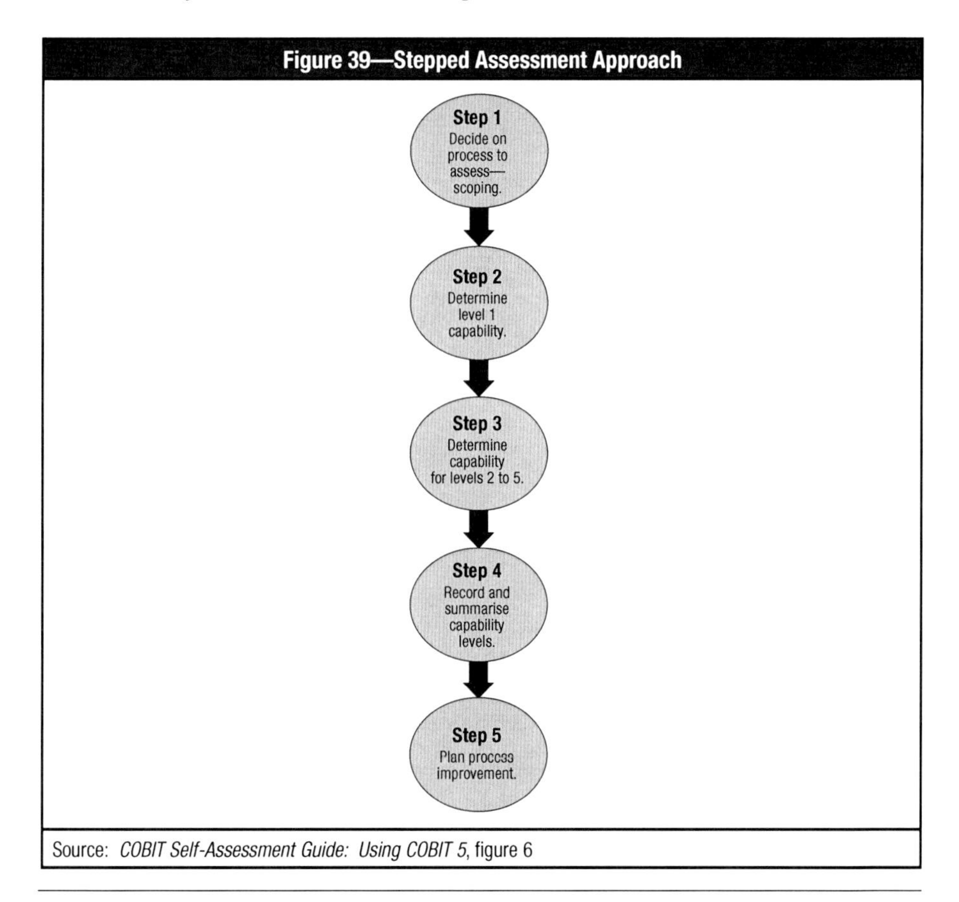

Source: *COBIT Self-Assessment Guide: Using COBIT 5*, figure 6

Step 1—Scope of the Example Assessment

Based on the assurance initiative, as performed in appendix B, six key processes have been identified in scope of the cloud computing assurance program. A capability assessment can be performed for each of these six processes. By means of example, this chapter focus is on the process APO10 *Manage suppliers*, specifically, Manage Cloud Service Providers (CSPs), across all of the phases (evaluate, implement, manage and monitor). Each of the other key processes can be assessed in a similar way, leading to a capability level for each process. Finally, a capability level can be established for the whole of cloud computing in an enterprise, based on each of the capability levels of the key processes.

Steps 2 and 3—Example Assessment

1. **Performed process**—The first level focuses on the process achieving its goals. Although defining target capability levels is up to each enterprise to decide, many enterprises will have the ambition to have all of their processes achieve capability level 1. Assessing whether the process achieves its goals—achieves capability level 1—can be done by verifying some basic steps of the APO10 process:
   - Are preferred and alternative CSPs identified, recorded and categorized?
   - Are CSP-specific evaluation criteria established to evaluate both the contract and performance of the CSP?
   - Is the selection process of a CSP documented and does it show clear focus on the evaluation of specific risk that outsourcing services to the cloud represents? (See chapter 1.)
   - Is the performance of the existing CSPs evaluated on a regular basis and compared to alternative CSPs?
   - Does the contract with the CSP contain clauses that are focused on specific cloud risk, e.g., data location, data security, privacy, network security, and thus includes cloud-specific SLAs? For example, a CSP was selected from a list of possible CSPs after undergoing an evaluation, and an adequate product/service was selected, based on the experience of the project manager, because there was no standardized approach available within the enterprise.

   When reviewing these process outcomes, not every outcome will be fully achieved. To have some sort of classification, an ISO/IEC 15504 rating scale can be used to assign a rating to what degree each objective is achieved. This scale consists of the following ratings:
   - **N (Not achieved)**—There is little or no evidence of achievement of the defined attribute in the assessed process. (0 to 15 percent achievement)
   - **P (Partially achieved)**—There is some evidence of an approach to, and some achievement of, the defined attribute in the assessed process. Some aspects of achievement of the attribute may be unpredictable. (15 to 50 percent achievement)
   - **L (Largely achieved)**—There is evidence of a systematic approach to, and significant achievement of, the defined attribute in the assessed process. Some weakness related to this attribute may exist in the assessed process. (50 to 85 percent achievement)

- **F (Fully achieved)**—There is evidence of a complete and systematic approach to, and full achievement of, the defined attribute in the assessed process. No significant weaknesses related to this attribute exist in the assessed process. (85 to 100 percent achievement)

In addition, the process (governance or management) practices can be assessed using the same rating scale, expressing the extent to which the base practices are applied. If this first capability level is not achieved, the reasons for not achieving this level are immediately obvious from the approach explained previously, and an improvement plan can be defined:

- If a required process outcome is not consistently achieved, the process does not meet its objective and needs to be improved.
- The assessment of the process practices will reveal which practices are lacking or failing, enabling implementation and/or improvement of those practices to take place and allowing all process outcomes to be achieved.

2. **Managed process**—The following examples of the COBIT 5 enablers can be expected at a "managed' capability level, covering the process attributes PA 2.1 Performance Management and PA 2.2 Work Product Management:
   - **Organisational Structures:**
     1. The process documentation sets up clear and defined roles, defining ownership, responsibility and accountability for the process of managing CSPs.
     2. The process documentation also includes the individuals and groups that will be consulted during the process. It clearly states at what moment of time in the process the roles are involved (e.g., during contract evaluation, performance evaluation) and what their specific assignment and specialty is in the process.
   - **Information:**
     1. The process plan sets clear performance targets for the process of managing CSPs and provides details on actions to take when performance is not achieved.
     2. A quality plan is established and provides details on the quality criteria of the contracts and evaluations of the CSPs. The quality criteria take into account the specificities of cloud computing, e.g., the flexibility that cloud computing provides, which should be reflected in the contract.
     3. A quality record is established to provide an audit trail of reviews undertaken. This audit trail shows that people who are knowledgeable on cloud computing were involved in the process.
   - **People, Skills and Competencies:**
     1. The designated roles in the process clearly demonstrate having the required cloud computing knowledge to assist in the process (e.g., specific cloud computing knowledge will be needed to determine the service levels in the SLAs with the CSP).

For example, in this situation, a CSP was selected by the vendor management department, using a standardized approach and predefined documents and templates. Supplier risk was also taken into account. The efficiency of the department is evaluated on a yearly basis.

3. **Established process**—The following examples of the COBIT 5 enablers can be expected at an "established" capability level, covering the process attributes PA 3.1 Process Definition and PA 3.2 Process Deployment:
   - **Principles, Policies and Frameworks:**
     1. A defined CSP management process model is established to enable the execution of the process in a repetitive manner in such a way that it will always achieve its process outcomes. Policies and standards provide details on the organizational objectives of the process, expected sequences and minimum performance standards. The requirement here is not that policies and standards exist, but that they are actually applied.
   - **Processes:**
     1. The sequence and interaction with other disciplines is determined.
   - **Information:**
     1. Data are being collected and analyzed as a basis for understanding the behavior of, and to demonstrate the suitability and effectiveness of, the process, and to evaluate where continuous improvement of the process can be made.
   - **Services, Infrastructure and Applications:**
     1. The supporting application(s) for the evaluation and management of a CSP is (are) identified, deployed and maintained. All CSP contracts, SLAs, and evaluations are stored in and managed via this tool.
   - **People, Skills and Competencies:**
     1. The required competencies to perform the defined CSP management process are identified.
     2. People involved in the process have received appropriate education, training and experience to manage CSPs. They have been trained in understanding the specificities of cloud computing and are thus capable of managing a CSP.

   For example, in this situation a CSP was selected by the vendor management department, using the purpose-built vendor selection tool which enforces the established process flow and provides regular reporting. The process is well documented and communicated throughout the enterprise. Proper training on the process and the tooling is being provided for all relevant staff members.

4. **Predictable process**—The following examples of the COBIT 5 enablers can be expected at a "predictable" capability level, covering the process attributes PA 4.1 Process Measurement and PA 4.2 Process Control:
   - **Processes:** A process measurement plan is defined and contains:
     1. Quantitative objectives for the CSP management process performance, in support of business goals
     2. Control limits of variation for normal CSP management process performance.
     3. Identified measures and frequency of measures, in line with the CSP management process measurement needs.

   Subsequently,
   - Results of measures are collected, analyzed and reported.
   - Corrective action is taken to address variation.
   - Control limits are re-established following corrective action.

- **Information:**
  1. A process improvement plan is established and provides improvement objectives and proposed improvement actions.
  2. Process performance records are being maintained to track details of measurements collected. Both information items are interconnected with the actions noted under "Processes."

For example, in this situation, the vendor management department is compliant with the previous level and specific KPIs on performance and completeness have been established and are closely monitored for noncompliance. Noncompliance results in corrective actions.

5. **Optimising process**—The following examples of the COBIT 5 enablers can be expected at an "optimized" capability level, covering the process attributes PA 5.1 Process Innovation and PA 5.2 Process Optimization:
   - **Organisational Structures:**
     1. The structured organization (including roles and responsibilities) to manage the organization's CSPs is regularly assessed in terms of suitability in accordance to the (changed) business environment and their needs.
   - **Processes:**
     1. The CSP management process is regularly assessed in terms of suitability in accordance to the (changed) business environment, their needs and also to the rapidly changing world of cloud computing itself.
     2. The CSP management process is also regularly assessed to identify opportunities for best practice and innovation.
   - **Information**—The process improvement plan:
     1. Includes identified improvement opportunities from new technologies and innovations
     2. Establishes an implementation strategy for the process improvement
     3. Evaluates the effectiveness of process change based on actual performance against the defined process objectives
     4. Furthermore, evidence of the improvement to the process should be visible in the process documentation, quality plan, and related policies and standards.

   For example, in this situation, the vendor management department is compliant with the previous level and there is a strict alignment with business goals, needs and strategy.

Steps 4 and 5—Assessment Maturity vs. Target Maturity

A spider graph can be used to visually display the capability assessment results. **Figure 40** shows an example of a process capability assessment and displays both the target capability level and the assessed capability level for the APO10 process.

Based on this assessment between current and target maturity, a process improvement plan (step 5) can be set up and executed.

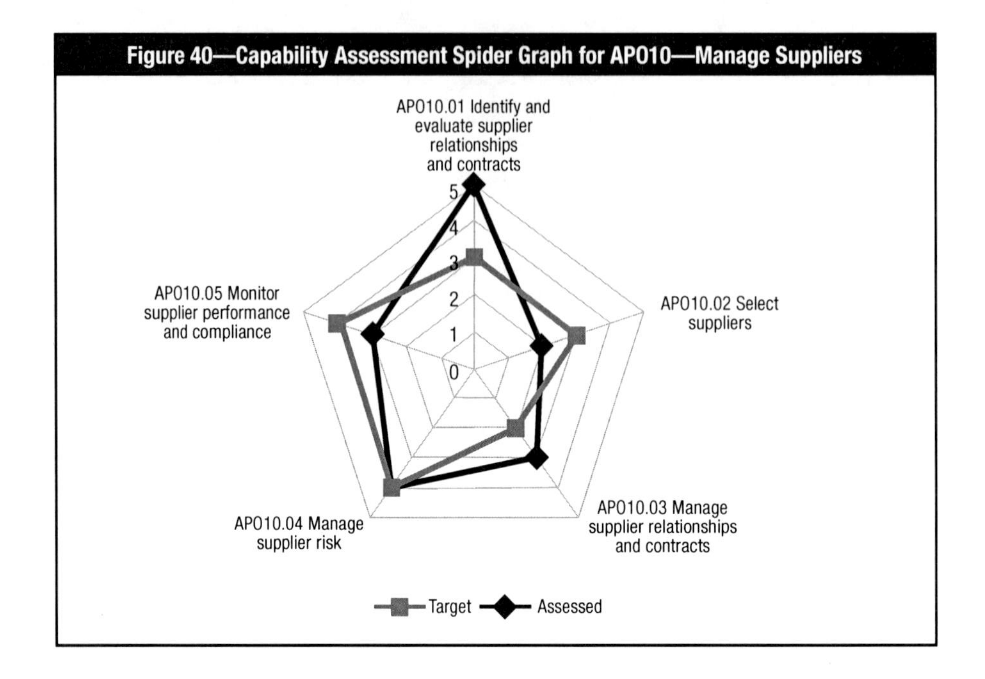

Figure 40—Capability Assessment Spider Graph for APO10—Manage Suppliers
APO10.01 Identify and evaluate supplier relationships and contracts
5
4
3
2
1
0
APO10.02 Select suppliers
APO10.03 Manage supplier relationships and contracts
APO10.04 Manage supplier risk
APO10.05 Monitor supplier performance and compliance
Target
Assessed

Appendix D. Cloud Risk Scenarios

A risk scenario is a description of a possible event that, when occurring, will have an uncertain impact on the achievement of the enterprise's objectives. The impact can be positive or negative.

The COBIT 5 risk management process requires risk to be identified, analyzed and acted on. Well-developed risk scenarios support these activities and make them realistic and relevant to the enterprise.

Risk scenarios can be derived via two different mechanisms:

- A top-down approach, where one starts from the overall enterprise objectives and performs an analysis of the most relevant and probable IT risk scenarios impacting the enterprise objectives. If the impact criteria used during risk analysis are well aligned with the real value drivers of the enterprise, relevant risk scenarios will be developed.
- A bottom-up approach, where a list of generic scenarios is used to define a set of more relevant and customized scenarios, applied to the individual enterprise situation.

The approaches are complementary and should be used simultaneously. Risk scenarios must be relevant and linked to real business risk. Also, using a set of example generic risk scenarios could assist to identify risk and reduce the chance of overlooking major/common risk scenarios and can provide a comprehensive reference for IT risk. However, specific risk for each organization and critical business requirements will need to be considered in the organization's risk scenarios.

More information on how to develop risk scenarios can be found in *COBIT 5 for Risk.*

Cloud Risk Scenario Examples

Figure 41 lists a series of positive and negative example scenarios for each of the generic cloud risk scenario categories. The related risk type is always indicated via a "P" (Primary) or an "S" (Secondary). This allows for a structured way to identify, analyze and respond to risk.

Figure 41—Example Cloud Risk Scenarios					
	Risk Type			Example Scenarios	
Risk Scenario Category	IT Benefit/Value Enablement	IT Program and Project Delivery	IT Operations and Service Delivery	Positive Example Scenarios	Negative Example Scenarios
New (cloud) technologies	P	S	P	• New technologies for new initiatives or more efficient operations adopted and exploited • Competitive advantage • Business innovation potential	• Failure to adopt and exploit new technologies (i.e., functionality, optimization) on a timely basis • New and important technology trends not identified • Inability to use the technology to realize desired outcomes (e.g., failure to make required business model or organizational changes)
(Cloud) technology selection	P	S	S	• Optimal technology selected for implementation • Ability to switch faster to newer technology	Wrong technologies (i.e., cost, performance, features, compatibility) selected for implementation
Cloud investment decision making	P	S		• Coordinated decision making over IT investments between business and IT • Reduced IT investment	Business managers or representatives not involved in important IT investment decision making (e.g., new applications, prioritization, technology opportunities)
Accountability over cloud (IT)	P	S	S	Business assumes appropriate accountability over IT and codetermines the strategy for the cloud, especially application portfolio	Business not assuming accountability over those cloud areas for which it should (e.g., functional requirements, development priorities, opportunity assessment through new technologies)
Integration of cloud computing with business processes	S		P	Fully integrated cloud solutions in place across business processes	Separate and nonintegrated solutions to support business processes

Figure 41—Example Cloud Risk Scenarios *(cont.)*					
	Risk Type			Example Scenarios	
Risk Scenario Category	IT Benefit/Value Enablement	IT Program and Project Delivery	IT Operations and Service Delivery	Positive Example Scenarios	Negative Example Scenarios
Integration with legacy systems	S		P	Systems with interoperability that work together to effectively handle enterprise information	Systems unable to work together and key information not accessible. Security issues and duplication of efforts may also arise, resulting in increased costs for the enterprise.
State of cloud (infrastructure) technology	P		P	Modern and stable technology used	Obsolete technology in use that cannot satisfy new business requirements (e.g., security, storage)
Architectural agility and flexibility	S	S	P	Modern and flexible architecture that supports business agility/innovation	Complex and inflexible IT architecture obstructing further evolution and expansion
Software implementation in the cloud		P	S	Faster development and higher quality of testing	• Operational glitches when new software is made operational • Users not prepared to use and exploit new application software
Selection/ performance of cloud suppliers	S	P	S	• Cloud supplier acting as strategic partner • More choices and information available on cloud suppliers	• Inadequate support and services delivered by vendors, not in line with SLAs • Inadequate performance of CSP in large-scale, long-term cloud arrangements
Contractual compliance	S		P	CSP exceeds contractual obligations	Contractual obligations by CSP with customers/clients not met

Page intentionally left blank

Appendix E. Contractual Provisions

This appendix provides a short overview of important contractual provisions for custody and ownership; security, privacy and access; and business continuity, data disposal and exit strategy that an enterprise should consider when contracting with a CSP.

Scope of Services

The IT contract must clearly define the services that will be performed, to avoid any dispute among the parties regarding scope of services. For a time and materials (T&M) contract, it is especially important to clearly describe the scope of the services to be provided, the scheduling of the services and the number of services to be delivered. Alternatively, when the IT contract is for a well-defined deliverable (e.g., specific software that will be developed), establishing the scope becomes less important as long as the deliverable specifications and requirements are clearly defined in the contract.

Special attention should be paid to defining the terminology used throughout the agreement, particularly terms that are commonly used in one field or location, but may not be known or may have another meaning in other fields or locations (e.g., best effort, commercially reasonable effort and *force majeure*).

IT projects may have a dynamic nature, so the contract may be subject to frequent changes, especially in the case of long-term agreements. For this type of project, establish a method to modify the scope, i.e., a change management procedure, or include a consultation clause in the agreement. This allows the parties to distinguish the "must have" and the "nice to have," assess risk associated with proposed changes while the project progresses, and jointly approve changes.

Core Obligations of Parties

After the scope of services has been clearly defined, parties should agree on the performance standards regarding the delivery of services.

IT contracts usually contain a description of mutual rights and obligations for the parties. Most often, these obligations consist of general responsibilities, such as information obligation, cooperation obligation and obligation to comply with the applicable legislation.

The parties should set sufficient interim delivery milestones to enable the project to be completed successfully and in due time. Such milestones allow the enterprise to monitor the project progress and to maintain control of the various interim deliverables. Each milestone and the corresponding interim deliverables should be clearly defined by the parties in the contract.

End of Contract

The way in which a contract is terminated depends on the term for which it has been concluded (definite or indefinite) and on the termination possibilities for the parties included in the contract. Including specific exit arrangements in the IT contract is important. A smooth transfer to another service provider or in-house prevents endangering the continuity of business operations. Ownership rights and the right to use intellectual property after the end of the contract, are part of the end of contract arrangements.

Due to the technical complexity of IT projects, some IT contracts consist of several independent contracts and appendices. As a consequence, the termination of one contract may entail the termination of another contract (e.g., a maintenance contract). IT contracts need to be properly monitored to ensure that all dependencies are properly addressed during termination procedures.

Custody and Ownership

Contractual provisions should:
- Explicitly state that the enterprise is the owner of all rights, title and interest in the data and that all data will be maintained, backed up and secured until returned on termination of the agreement (unless other provisions are made for the migration, transfer or destruction of the data).
- Identify the geographic locations where data storage and processing will occur.
- Confirm the jurisdiction that governs the operation of the contract and the application of privacy, confidentiality, access and information management laws.
- Confine data storage and processing to specified locations where the regulatory framework and technical infrastructure allow the enterprise to maintain adequate control over the data.

Security, Privacy and Access

Contractual provisions to consider:
- Specify security standards with which the provider must demonstrate compliance, including a warranty in relation to security, related storage and access obligations and SLAs that include the cost and operating requirements of providing service continuation in business-critical and nonbusiness-critical services when disruptions arise.
- Prescribe the security provisions that the service provider must implement that are consistent with the enterprise's digital information security policy; for example, where required, certified compliance with AS/NZS ISO/IEC 27001 Information technology—Security techniques—Information security management systems, or equivalent standards.
- Prohibit any unauthorized access, use or alteration of the data. Document the technical mechanisms and procedures in place to support this restriction, e.g., enterprise control of user credentials for authentication, data encryption, information

dispersal, data separation and segregation. Ensure that the contract prevents unauthorized access or use by the service provider or subcontractor, including any third-party use of data.
- Investigate and guarantee by contractual terms that the computing processes by which the CSP secures its information and the encryption between the enterprise and any overseas cloud storage location.
- Allow enterprises to receive data breach notifications.
- Include terms of the provider's information security management system.
- Specify that that any personal information that is contained in the data is subject to Privacy and Personal Information Protection Act (PPIPA), Health Records and Information Privacy Act (HRIPA) and any other relevant legislation, as applicable. Include the specification that any person or body providing data services relating to the collection, processing, disclosure or use of personal information for, or on behalf of, a public sector agency must abide by the information protection principles.
- State that the enterprise retains an immediate and ongoing right of access to all organizational data held by the CSP.
- Include provisions that allow the data or the service to be audited to ensure that enterprises meet their requirements under policy and legislative frameworks.

Fees and Payment Terms and Modalities

The contract should be explicit about the method to calculate price. It should be clearly specified whether the project will be executed for a fixed price, on the basis of actual time spent (e.g., an hourly rate or a man per day rate) or a combination of both.

It is equally important for the payment terms and modalities to be included in the contract (e.g., send an invoice, the term of payment and the consequences of late payment). For long-term contracts, it is advisable to include an adjustment clause, which allows parties to modify certain conditions (e.g., the price) of the agreement at specific stages of the project. Depending on the applicable legislation, an indexation clause could be inserted into the contract. This clause should contain, at a minimum, the index, the indexation formula and the frequency of the indexation. Other price revision methods can also be included (e.g., consultation clause).

Performance

An SLA is very common in IT contracts and can be an effective tool to create a common understanding between contracting parties regarding services, priorities and/or responsibilities in the framework of an agreement.

Only measurable and realistic SLAs are defined in the contract. Unrealistic SLAs might cause disputes between parties. SLAs should be measurable to allow meaningful reporting among the parties. The reports can be used during consultations between parties about the achievement of the SLA.

It is important to address the consequences of noncompliance in the SLA. The contract could provide a set of actions to deal with the cause of noncompliance. The parties could define penalties in case of nonachievement of the levels established. However, incentives for the achievement of the SLA (or a combination of penalties and incentives) often appear to be more efficient.

If the parties decide to provide a price correction, it is important to stipulate that the price correction is not the only compensation that can be claimed in a case of nonachievement of the SLA, so that the right to the payment of full damage is not jeopardized. For continuity of the enterprise, it is recommended to provide a special clause that states that the enterprise can appoint a third-party service provider to deliver the necessary services in case of nonachievement of the SLA by the service provider.

If required, an audit clause can be included in the contract that allows the enterprise to obtain the right to access relevant financial data, accounting information, operational and technical data, safety standards, etc., of the other party. It is strongly recommended to insert this audit clause into a contract because it allows the enterprise to check, among others, the co-contracting party's performances, compliance and achievements.

Term and Termination Possibilities

Contracts can be established for a definite or indefinite term. A contract for an indefinite term can be terminated at any moment in time, subject to due notice. A contract for a definite term ends automatically upon expiration of its term. Nevertheless, parties could agree to renew the contract automatically (tacit agreement). It is important to define in the contract the consequences of the automatic renewal (e.g., will the same terms and conditions apply, what will be the duration of the renewed contract and will the same SLAs be in place).

If the parties foresee the possibility of interim termination due to specific circumstances, they could include a clause that allows early termination of the contract for specific reasons, without any court intervention, and thus avoid long judicial procedures. For example, the parties could include a material adverse change clause (MAC clause), which allows a contracting party to terminate or renegotiate the contract in case of a substantial (material) change in the circumstances (e.g., in case of a change to the current legislation). MAC clauses are especially useful in long-term agreements. The clause should identify in a concrete and objective manner the different circumstances that can lead to an early termination or renegotiation of the agreement and the possible consequences in the case of the specified material adverse change (e.g., price adjustments, termination of the contract or renegotiation of the contract delivery terms, and product specification).

The parties can define which circumstances or events of default will lead to an early dissolution of the agreement with entitlement to compensation. As indicated previously, proper exit arrangements should be negotiated to ensure a smooth transfer to another provider in case of termination under any circumstances.

Liabilities

Dependent on the jurisdictions involved, the legal rules regarding liability are quite stringent. However, contracting parties might be able to negotiate different arrangements regarding liabilities. The service provider, in particular, will benefit from a limitation of liability because the service provider bears the biggest responsibilities. The limitation of the liability of the service provider regarding the security, protection and loss of data is a common liability clause in contracts. In most cases, the service provider will try to limit its liability to the minimum extent possible.

Contractual clauses regarding the limitation of liability can take several forms, for example, the liability can be capped to a maximum amount or certain damages can be excluded from any compensation. It is common practice for a service provider to try to limit its liability as much as possible. However, the service provider should bear some liability, especially regarding its core obligations. Furthermore, such limitation of liability should be evaluated in view of the enterprise's liability toward its own customers as well as the liabilities covered by insurance policies.

Parties can also define how third-party claims will be addressed (e.g., in case of an infringement on intellectual property rights). Include indemnity clauses in the contract that define how these claims will be handled. When a contracting party of the enterprise bears damages because of a fault of the service provider, the enterprise will be held responsible by the contracting party. Because of the indemnity clause, the enterprise can seek to have the service provider reimburse it for monies that the enterprise had to pay to the damaged party.

Certain jurisdictions have special liability laws for online providers (e.g., laws of the member states of the European Union [EU] and the Digital Millennium Copyright Act of the USA).

Dispute Resolution and Applicable Law

The IT services contract may involve different countries and, therefore, cover different jurisdictions; for example, the service provider and the enterprise are located in different countries. The contract must determine the governing law and the competent jurisdiction in case a dispute arises related to the agreement. The choice of the applicable law has an impact on the validity and enforceability of the clauses of the agreement. A legal procedure under foreign law or in front of a foreign court may involve additional costs (e.g., the intervention of a foreign lawyer) and risk.

Business Continuity, Data Disposal and Exit Strategy

Contractual provisions to consider:

- Document the technical mechanisms and procedures that prevent data loss; e.g., contractor/enterprise responsibilities and routines for backup, failover or redundancy.

- Provide for continuity of accessibility, usability and preservation of all agency data, regardless of any migration of data to other formats during the contract. Terms should provide for appropriate testing to ensure data integrity prior to any migration.
- Specify provisions and procedures for backup, restoration of services and disaster recovery.
- Upon transfer of data, ensure that technological parity with other service providers is guaranteed.
- Guarantee the preservation of data and provide for routine monitoring of data to identify formats that are at risk of obsolescence.
- Include provisions relating to the migration of data to new formats, when appropriate, and the provision of proper documentation about migration activities to the enterprise.
- Include provisions for the safe return/transfer of data if the CSP is the subject of a takeover.
- At the termination of the agreement with a CSP, specify what will happen to the data; for example, transfer to a new provider, return to the enterprise or permanently delete.
- Specify remedies for service provider mistakes or breaches.
- Identify penalty provisions imposed by the service provider; e.g., suspension of enterprise access to the data because of nonpayment.
- Define contract provisions relating to the migration of data on termination of the contract.
- Precisely specify the terms for the disposal of specific data during the term, at the request of the enterprise and at the end of the term, including a warranty in relation to technological parity/obsolescence.
- Limit suspension and termination rights that are available to the CSP.
- Allow subscription levels to be scalable up and down according to demand.
- Specify reporting and audit rights of the enterprise and vendor.

Appendix F. Cloud Enterprise Risk Management (ERM) Governance Checklist

An enterprise risk management (ERM) governance checklist is a baseline for developing more in-depth communications within public and private enterprises, because both encounter different regulatory obligations, business models, an aversion to risk and the willingness to aggressively guard against fraud, both internally and externally. The governance and regulatory requirements will impact the executive boardroom's duties and responsibilities regarding all aspects that are associated with conducting their fiduciary due diligence prior to selecting the appropriate cloud solution for their business.

Cloud Enterprise Risk Management (CERM) Governance Checklist

Figure 42 is a list of cloud-related questions that an enterprise's board of directors should consider asking in its governance oversight role.

Figure 42—Cloud-related Questions for the Board of Directors to Consider

Cloud Enterprise Risk Management Topic	Comment/ Response	Responsible Role/Name	Last Update Date	Last Review Date
What level of consideration has management given to adopting cloud computing, and what is management's current position on this area?				
Who in management is responsible for understanding and managing the business risk associated with cloud computing?				
What are competitors doing with cloud computing solutions?				
Does management have effective processes in place to monitor cloud computing adoption and usage?				
What would be the impact of cloud computing to management's overall internal control structure (improved, unchanged or diminished)?				
Does management have the skills required to understand the complexities associated with cloud computing?				
Are cloud computing initiatives aligned with the enterprise's risk appetite?				

Figure 42—Cloud-related Questions for the Board of Directors to Consider *(cont.)*

Cloud Enterprise Risk Management Topic	Comment/ Response	Responsible Role/Name	Last Update Date	Last Review Date
Are due diligence processes adequate for assessing cloud computing vendors at both the initial contract stage and the engaged stage, which requires monitoring processes?				
Has management established adequate, minimum, service-level expectations for third-party CSPs?				
How is management mitigating organizational risk resulting from reliance on the activities of a third-party CSP?				
If cloud computing solutions are being used to support the enterprise, has the cloud computing risk been determined and disclosed to investors (where applicable)?				
What is management's stance on outsourcing, and does this stance align with the current approach of the enterprise to cloud computing?				
Does the enterprise anticipate rapid growth that might require using cloud solutions?				
Is the enterprise in a mature market that might require using cloud computing to save costs to remain competitive?				
Are the enterprise's operational functions and processes mature and formalized enough to allow for a change in the underlying technology platform?				
What is the capability and maturity of the enterprise's current IT function?				
How should the enterprise prepare for cloud computing?				
Should cloud computing be embraced, to capitalize on its benefits, or rejected, to avoid risk such as data breaches or noncompliance with complex e-discovery requirements?				
How can the enterprise manage its risk adequately, while operating in a business environment with cloud computing?				

Appendix G. A Practical Approach to Measuring Cloud ROI

Figure 43 outlines the three phases and suggested questions to address each step.

Figure 43—Cloud Phases and Steps	
Phase/Step	**Guidance/Key Questions to Answer**
Phase 1—Determine To-be Cloud Costs and Benefits	
a. Define high-level business (functional) requirements.	• What business functions need to be covered? • What are the business drivers for adopting cloud-based services? • How might cloud-based services support business processes? • What compliance requirements (e.g., US Sarbanes-Oxley Act, HIPAA, PCI DSS) are relevant?
b. Define initial/baseline cloud service model.	• What type of cloud service model (IaaS, PaaS, or SaaS) is needed? • What type of deployment model (public, private, community or hybrid) is most appropriate? • Where would services be physically located (e.g., on premises, off premises, specific geographical location)? • Who would deliver the services (e.g., third party, in-house, mix, cloud broker)? • For this baseline, start with a model that is simple and low-cost (e.g., public SaaS), but rule out options that will not meet major compliance requirements (e.g., focus on in-country providers if use of foreign providers is prohibited). • The baseline solution may not be the optimal one or may fall outside the enterprise's risk tolerance, but later steps should address these concerns.
c. Risk-assess initial/baseline cloud model.	• Identify the risk areas to be considered (e.g., multitenancy, network dependency, abstraction, data usage limitations, security, privacy, up-front migration cost, cross-border data location, vendor lock-in, hardware lock-in, data ownership, in-house skills required to manage the cloud). • Determine countermeasures to mitigate the areas of risk outside of the enterprise's risk tolerance. • Examples of risk mitigation measures may include: – Data encryption/tokenization managed by the customer to protect against unauthorized data access by cloud provider staff – A revert-back strategy to protect against potential failure of the cloud provider's business – Backups/audit trailing housed on customer premises to protect against loss of access to cloud services – Clear and comprehensive SLAs that include the right to audit clause – Implementation of in-house DRP

Figure 43—Cloud Phases and Steps *(cont.)*	
Phase/Step	**Guidance/Key Questions to Answer**
Phase 1—Determine To-be Cloud Costs and Benefits ***(cont.)***	
d. Estimate costs.	The costs may include: • Cost of migrating from the current model to a cloud-based model (e.g., rewriting applications to operate in a virtualized environment, reformatting data to suit SaaS provider formats, setting up federated identity and access management, implementing processes to manage the cloud) • Cost of operating the cloud-based model (e.g., cloud provider fees, software licensing and support fees, data communication fees, cloud system administration) • Cost of implementing and operating countermeasures to mitigate risk (e.g., data encryption tools, planning and testing revert-back strategies, maintenance of backups and audit logs offline from provider) The calculations must include other factors as well: • Estimate tangible benefits (e.g., increased sales due to improved availability, scalability of systems, increased revenue from sales representatives having better access to information while traveling, reduced head count supporting traditional IT systems). • Assess intangible costs/benefits. These may include such considerations as: – Ability to react quickly to changing markets through rapid product release and/or scaling – Potential that cloud providers will be able to support introduction of new technical innovations faster than a traditional IT function could – Risk that tightening of regulations (e.g., privacy) may make cloud services nonviable in the future, forcing systems to return in-house – Loss of internal IT skills/knowledge that could otherwise be a strategic differentiator – Risk of being locked in to particular cloud providers/proprietary service models, potentially impeding future adoption of open standards based services as they emerge
e. Consider other cloud models.	• Would it be more cost-effective to change the cloud service/deployment models? For example: – Instead of a public cloud, would private, community or hybrid clouds remove the need for some of the security controls required for a public cloud? – Instead of SaaS, would PaaS or IaaS make it more cost-effective to mitigate some of the lock-in risk? • Test each of the key alternatives available against the baseline model to determine if there is a more optimal cloud model for the circumstances.
f. Reevaluate costs/benefits to align to optimal model.	• Once an optimal model is determined, update the costs and benefits (migration costs, operating costs, risk mitigation costs, intangible costs/benefits) to reflect this model.

Figure 43—Cloud Phases and Steps *(cont.)*	
Phase/Step	**Guidance/Key Questions to Answer**
Phase 2—Evaluate As-is Costs and Benefits	
a. Estimate as-is costs and benefits.	• Using the same definition of business requirements as utilized in phase 1, define the current service model to meet the same functional and compliance requirements.
b. Perform (or review if one already exists) a risk assessment of the current service model.	• Are there any risk areas that fall outside the enterprise's risk tolerance that need to be mitigated? For example: – The current system is locked in to a particular technology or provider, and moving to the cloud may require considerable time, effort and money. – The current system contains intellectual property that requires high levels of security and compartmentalization. – The cloud provider may not meet current service levels, leading to degradation in service to customers and loss of business. • Determine countermeasures/mitigations required to bring this risk to an acceptable level (e.g., use a private cloud to avoid multitenancy, assess vendor performance, assess vendor certifications and compliance profile). • To ensure an "apples to apples" comparison, review the risk areas considered in the to-be assessment to ensure they have all been considered in the as-is assessment, and *vice versa*.
c. Estimate costs/benefits.	These may include: • Ongoing operation/maintenance costs (TCO) • Risk mitigation cost • Intangible costs/benefits
Phase 3—Estimate ROI	
a. Compare as-is and to-be costs and benefits.	• The simplest way to do this is to prepare a table comparing the quantified costs and benefits for the as-is and to-be options over a period of up to five years. A longer period is not recommended due to the speed with which the IT industry changes. • For each year, calculate the net incremental cost/benefit of moving to the to-be cloud solution.
b. Calculate ROI.	• Several methods can be used. Engage the enterprise's finance team and apply the organizational standard. • If the enterprise does not have a standard, the simplest approach is to use the simple ROI calculation, supported by a simple NPV calculation. To calculate NPV, take the net cost/benefit for each year and discount it back to the present using an approved interest rate (i.e., the rate at which the organization borrows). This should result in the total cost/benefit in present-day value.
c. Factor in intangibles.	• Cloud computing initiatives can involve significant intangible benefits (e.g., increased ability to release new products rapidly) and costs (e.g., potential loss of internal IT technical skills). • If these intangibles cannot be quantified reliably, they need to be described as clearly as possible and included in the ROI assessment to ensure that the final decision is based on a holistic set of factors.

Formulas

Total Cost of Ownership (TCO)

Cost of ownership analysis is intended to uncover the lifetime costs of acquiring, operating and maintaining something (services or assets). TCO is useful to determine the difference between the purchasing price and the long-term cost of investment.

TCO Formula

$$TCO = Purchase + Financing + Maintenance + Upgrade + Enhancements + Deployment + Security + Depreciation + Decommissioning + Disposal + Cost_n$$

The period of time used to calculate TCO depends on corporate standards which determine when ownership starts and ends. Three common life spans are:

- Depreciable life
- Economic life
- Service

Net Present Value (NPV)

Net present value is intended to calculate the present value of an investment by the discounted sum of cash flow disbursements over a period of time.

NPV Formula

$$NPV = \text{Initial Investment} + \sum_{i=1}^{\text{Time}} \frac{\text{Cash Flow}_i}{(1 + \text{Discount Rate})^i}$$

Example of NPV for an investment of US $500,000 and a rate of 10 percent over 3 years:

Year	Cash Flow	Present Value
0	-$500,000	-$500,000
1	$200,000	$181,818
2	$300,000	$247,933
3	$200,000	$150,262
		Investment NPV = $80,015

$$NPV = -\$500,000 + \frac{\$200,000}{1.10} + \frac{\$300,000}{(1.10)^2} + \frac{\$200,000}{(1.10)^3}$$

Internal Rate of Return (IRR)

IRR is the interest rate that would make the initial investment NPV equal to zero. Usually, a rate greater than the cost of borrowing money would be considered beneficial by most finance professionals or portfolio managers. Please note that IRR cannot be derived analytically; instead, IRR must be found by using mathematical calculations to find the correct rate. However, most financial calculators or spreadsheet programs can be used to calculate IRR.

IRR Formula

$$0 = \text{Initial Investment} + \frac{\text{Cash Flow}_1}{(1 + \text{IRR})} + \frac{\text{Cash Flow}_2}{(1 + \text{IRR})^2} + \frac{\text{Cash Flow}_n}{(1 + \text{IRR})^n}$$

Example using the same investment of US $500,000 over 3 years and IRR of 19 percent (this is the rate that would make the initial investment zero).

Year	0	1	2	3
Cash Flow	-$500,000	$200,000	$300,000	$200,000
IRR = 19%				

$$0 = -\ \$500{,}000 + \frac{\$200{,}000}{1.19} + \frac{\$300{,}000}{(1.19)^2} + \frac{\$200{,}000}{(1.19)^3}$$

Page intentionally left blank

Glossary

Term	Definition
American Institute of Certified Public Accountants (AICPA)	US national professional organization of CPAs
Application programming interface (API)	A set of routines, protocols and tools referred to as "building blocks" used in business application software development **Scope note:** A good API makes it easier to develop a program by providing all the building blocks related to functional characteristics of an operating system that applications need to specify; for example, when interfacing with the operating system (OS) (e.g., provided by Microsoft Windows, different versions of UNIX). A programmer would utilize these APIs in developing applications that can operate effectively and efficiently on the platform chosen.
Application service provider (ASP)	Also known as managed service provider (MSP); deploys, hosts and manages access to a packaged application to multiple parties from a centrally managed facility **Scope note:** The applications are delivered over networks on a subscription basis.
Automated Audit, Assertion, Assessment and Assurance application programming interface (A6)	A cross-industry work group attempting to develop a common interface allowing cloud service providers (CSPs) to automate the audit, assertion, assessment and assurance of their cloud infrastructures
BITS (formerly stood for Banking Industry Technology Secretariat)	A nonprofit, chief executive officer (CEO)-driven financial service industry consortium made up of 100 of the largest financial institutions in the US. BITS works to sustain consumer confidence and trust by ensuring the security, privacy and integrity of financial transactions.
Business continuity plan (BCP)	A plan used by an enterprise to respond to disruption of critical business processes. Depends on the contingency plan for restoration of critical systems.
Business Model for Information Security (BMIS)	An ISACA holistic and business-oriented model that supports enterprise governance and management information security, and provides a common language for information security professionals and business management
Canadian Institute of Chartered Accountants (CICA)	Canadian national professional organization of CAs
Capital expense (CAPEX)	An expenditure that is recorded as an asset because it is expected to benefit more than the current period. The asset is then depreciated or amortised over the expected useful life of the asset.
Central processing unit (CPU)	Computer hardware that houses the electronic circuits that control/direct all operations of the computer system

Term	Definition
Change management (CM)	A holistic and proactive approach to managing the transition from a current to a desired organizational state, focusing specifically on the critical human or "soft" elements of change **Scope note:** Change management includes activities such as culture change (values, beliefs and attitudes), development of reward systems (measures and appropriate incentives), organizational design, stakeholder management, human resource policies and procedures, executive coaching, change leadership training, team building, and communications planning and execution.
Cloud governance committee	The function refers to the governance body that is charged with evaluating, directing and monitoring of the enterprise functions. The governance committee or the equivalent function that is charged with the governance of the enterprise often delegates responsibility for providing cloud governance to a specific cloud governance committee. Final accountability however stays with the governance committee.
Cloud Security Alliance (CSA)	A nonprofit organization designed to provide fundamental security principles to guide cloud vendors and to assist prospective cloud customers in assessing the overall security risk of a cloud provider
Cloud Service Broker (CSB)	A third-party provider with access to multiple datacenters and cloud service offerings. They integrate and tailor those various services to one service that can even be intertwined with enterprise in-house applications and systems.
Cloud service provider (CSP)	A provider of Software as a Service (SaaS), Platform as a Service (PaaS) and Infrastructure as a Service (IaaS) (SPI) services
COBIT® (formerly stood for Control Objectives for Information and related Technology)	A complete, internationally accepted framework for governing and managing enterprise IT that supports executives and management in their definition and achievement of business goals and related IT goals. COBIT describes five principles and seven enablers that support enterprises in the development, implementation, and continuous improvement and monitoring of good IT-related governance and management practices. **Scope note:** Earlier versions of COBIT focused on control objectives related to IT processes, management and control of IT processes and IT governance aspects. Adoption and use of the COBIT 5 framework is supported by guidance from a growing family of supporting products: • *COBIT 5: Enabling Processes* • *COBIT 5: Enabling Information* • *COBIT 5 Implementation* • *COBIT 5 for Information Security* • *COBIT 5 for Assurance* • *COBIT 5 for Risk* • *COBIT Assessment Programme* See *www.isaca.org/cobit* for more information

Term	Definition
International Organization for Standardization (ISO)	International Organization for Standardization (ISO) standards
Internet service provider (ISP)	A third party that provides individuals and enterprises access to the Internet and a variety of other Internet-related services
Intrusion detection	The process of monitoring the events occurring in a computer system or network to detect signs of unauthorized access or attack
Intrusion detection system (IDS)	Inspects network and host security activity to identify suspicious patterns that may indicate a network or system attack
Intrusion prevention system (IPS)	An extension of IDS that can prevent/block detection intrusions
IT Governance Institute® (ITGI®)	Founded by ISACA and its affiliated foundation in 1998; strives to assist enterprise leadership in ensuring long-term, sustainable enterprise success and increase stakeholder value by expanding awareness
IT Service Management Forum (itSMF)	itSMF is an independent, internationally, not-for-profit organization of IT Service Management (ITSM) professionals worldwide. They collect, develop and publish "best practice," support education and training, discuss the development of ITSM tools, initiate advisory ideas about ITSM and hold conventions. They are concerned with promoting ITIL (IT Infrastructure Library), Best Practice in IT Service Management and have a strong interest in the international ISO/IEC 20000 standard.
Japanese "Sarbanes- Oxley Act" (J-SOX)	The Japanese government's equivalent to the US Sarbanes-Oxley Act
Korean "Sarbanes- Oxley Act" (K-SOX)	The Korean government's equivalent to the US Sarbanes-Oxley Act
Logical partitioning (LPAR)	The IBM definition of dynamic partitioning, which is the variable allocation of central processing unit (CPU) processing and memory to multiple applications and data on a server.
National Institute for Standards and Technology (NIST)	Agency of the US Department of Commerce whose mission is to promote US innovation and industrial competitiveness by advancing measurement science, standards and technology *www.nist.gov/public_affairs/general_information.cfm*
Open Web Application Security Project (OWASP)	An open-source application security project. The OWASP community includes corporations, educational organizations and individuals from around the world.
Operating expense (OPEX)	An ongoing cost of performing daily business activity, e.g., utilities, insurance, maintenance, office supplies

Term	Definition
Operating system (OS)	A master control program that runs the computer and acts as a scheduler and traffic controller **Scope note:** The OS is the first program copied into the computer's memory after the computer is turned on and must reside in memory at all times. It is the software that interfaces between the computer hardware (disk, keyboard, mouse, network, modem, printer) and the application software (word processor, spread sheet, email), and it also controls access to the devices, is partially responsible for security components and sets the standards for the application programs that run in it.
Organization for the Advancement of Structured Information Standards (OASIS)	A nonprofit consortium that drives the development, convergence and adoption of open standards for the global information society
Payment Card Industry Data Security Standards (PCI DSS)	Worldwide information security standards defined by the Payment Card Industry Security Standards Council
Personal computer (PC)	An electronic data processor for use by an individual
Personal digital assistant (PDA)	Handheld devices that provide computing, Internet, networking and telephone characteristics; also known as "palmtop" and "pocket computer"
Personal Information Protection and Electronic Documents Act (PIPEDA)	Canadian law relating to data privacy
Personally identifiable information (PII)	Information that can be used alone or with other sources to uniquely identify, contact or locate a single individual
Platform as a Service (PaaS)	Offers the capability to deploy onto the cloud infrastructure customer-created or -acquired applications that are created using programming languages and tools supported by the provider
Project Management Body of Knowledge (PMBOK)	Developed by the PMI, this is an American National Standards Institute (ANSI)-norm for project management and aspires to describe and standardize well-known and widely used project management insights and methodologies.
Project Management Institute (PMI)	PMI is the world's largest not-for-profit membership association for the project management profession. They own globally recognized standards and certification program (e.g., PMP, PgMP, PfMP, PMI-ACP, PMI-RMP, PMBOK), extensive academic and market research programs, chapters and communities of practice, and professional development opportunities.
Projects In Controlled Environments, version 2 (PRINCE2)	PRINCE2 is a project management methodology. It was developed by the UK government agency Office of Government Commerce (OGC). The methodology encompasses the management, control and organization of a project.
Recovery point objective (RPO)	Determined based on the acceptable data loss in case of a disruption of operations. It indicates the earliest point in time to which it is acceptable to recover the data. The RPO effectively quantifies the permissible amount of data loss in case of interruption.

Term	Definition
Recovery time objective (RTO)	The amount of time allowed for the recovery of a business function or resource after a disaster occurs
Remote file inclusion	A type of vulnerability most often found on web sites, allowing an attacker to include a remote file usually through a script on the web server
Request for information (RFI)	Process to collect written information about the capabilities of various suppliers
Request for proposal (RFP)	Solicitation made, often through a bidding process, by an agency or company interested in procurement of a (cloud) service, to potential suppliers to submit business proposals
Sarbanes-Oxley Act of 2002 (SOX)	A US law enacted as a reaction to a number of major corporate and accounting scandals. Its intent is to ensure that publicly traded companies in the US maintain open transparency in their accounting procedures and have controls in place to prevent any manipulation of financial data.
Security incident and event management (SIEM)	A prepared defense plan to manage and document an enterprise's response to a security incident
Segregation/separation of duties (SoD)	A basic internal control that prevents or detects errors and irregularities by assigning responsibility for initiating and recording transactions and custody of assets to separate individuals **Scope note:** SoD is commonly used in large IT organizations so that no single person is in a position to introduce fraudulent or malicious code without detection.
Service level agreement (SLA)	An agreement, preferably documented, between a service provider and the customer(s)/user(s) that defines minimum performance targets for a service and how they will be measured
Service-oriented architecture (SOA)	A cloud-based library of proven, functional software applets that are able to be connected together to become a useful online application
Software as a Service (SaaS)	Offers the capability to use the provider's applications running on cloud infrastructure. The applications are accessible from various client devices through a thin client interface such as a web browser (e.g., web-based email).
Software as a Service, Platform as a Service and Infrastructure as a Service (SPI)	The acronym used to refer to the three cloud delivery models
Software/system development life cycle (SDLC)	The phases deployed in the development or acquisition of a software system **Scope note:** This is an approach used to plan, design, develop, test and implement an application system or a major modification to an application system. Typical phases of the SDLC include feasibility study, requirements study, requirements definition, detailed design, programming, testing, installation and postimplementation review, but not the service delivery or benefits realization activities.

Term	Definition
Storage Networking Industry Association (SNIA)	International organization for storing information
Structured Query Language (SQL)	A database computer language designed for managing data in relational database management systems (RDBMSs)
Total cost of ownership (TCO)	Includes original cost of the computer and software, hardware and software upgrades, maintenance, technical support, training, and certain activities performed by users
Uniform resource locator (URL)	Specifies where an identified online resource is available and the mechanism for retrieving it
Virtual private network (VPN)	A secure private network that uses the public telecommunications infrastructure to transmit data **Scope note:** In contrast to a much more expensive system of owned or leased lines that can only be used by one company, VPNs are used by enterprises for both extranets and wide areas of intranets. Using encryption and authentication, a VPN encrypts all data that pass between two Internet points, maintaining privacy and security
Virtualization	The process of adding a "guest application" and data onto a "virtual server," recognizing that the guest application will ultimately part company from this physical server

References

Australian Government Cloud Computing Policy, "Maximising the Value of Cloud," Commonwealth of Australia, 2013, *www.finance.gov.au/files/2012/04/Australian-Government-Cloud-Computing-Policy-Version-2.0.pdf*

British Computing Society, *Cloud Computing and Virtualisation: Two Years On*, Great Britain, January 2013

Cloud Security Alliance (CSA), Cloud Controls Matrix, USA, 2013, *cloudsecurityalliance.org/research/ccm*

ENISA:
- "Cloud Computing: Benefits, Risks and Recommendations for Information Security," 2009, *www.enisa.europa.eu/activities/risk-management/files/deliverables/cloud-computing-risk-assessment*
- "Cloud Computing Information Assurance Framework," 2009, *www.enisa.europa.eu/activities/risk-management/files/deliverables/cloud-computing-information-assurance-framework*
- "Critical Cloud Computing," 2012, *www.enisa.europa.eu/activities/Resilience-and-CIIP/cloud-computing/critical-cloud-computing*
- "Procure Secure: A Guide to Monitoring of Security Service Levels in Cloud Contracts," 2012, *www.enisa.europa.eu/activities/Resilience-and-CIIP/cloud-computing/procure-secure-a-guide-to-monitoring-of-security-service-levels-in-cloud-contracts*

IT Governance Institute (ITGI), *Board Briefing on IT Governance, 2nd Edition*, USA, 2003

ISACA:
- Auditing Security Risks in Virtual IT Systems, *ISACA Journal*, 2011 v1
- COBIT, *www.isaca.org/cobit*:
  - COBIT 5
  - *COBIT 5: Enabling Processes*
  - *COBIT 5 for Assurance*
  - *COBIT 5 for Information Security*
  - *COBIT 5 for Risk*
  - *COBIT 5 Implementation*
  - *COBIT 5 Online*
  - *COBIT 5 Process Assessment Model*
- Cloud resources, *www.isaca.org/Knowledge-Center/Research/Pages/Cloud.aspx:*
  - *Calculating Cloud ROI: From the Customer Perspective*
  - *Cloud Governance: Questions Boards of Directors Need to Ask*
  - *Security Considerations for Cloud Computing, www.isaca.org/cloud-security*
- *Vendor Management: Using COBIT 5, www.isaca.org/vendor-management*

Kanellos, Michael; "Is Cyber Monday Really Energy Efficient?," Greentech Enterprise, 24 November 2010, *www.greentechmedia.com/articles/read/is-cyber-monday-really-energy-efficient*

Mell, Peter; Timothy Grance; US National Institute of Standards and Technology (NIST) Special Publication (SP) 800-145 (Draft), "The NIST Definition of Cloud Computing," NIST, USA, 2011

New South Wales Government, "Cloud Services Policy and Guidelines," August 2013, *www.finance.nsw.gov.au/ict/sites/default/files/Endorsed%20-%20Cloud%20 Services%20Policy%20and%20Guidelines%20final%204%20Oct%20amend.pdf*

Pijanowski, Keith; "Understanding Public Clouds: IaaS, PaaS and SaaS," Keith Pijanowski's Blog, 31 May 2009

Stroud, Robert, "Providing Governance in a Rapidly Changing World," ISACA Euro Computer Audit, Control and Security (EuroCACS) conference 2010, Budapest, Hungary

US General Services Administration (GSA), "Cloud-based Infrastructure as a Service Comes to Government," 19 October 2010, *www.gsa.gov/portal/content/193441*